Urban Responses to Climate Change

Framework for Decisionmaking and Supporting Indicators

Debra Knopman, Robert J. Lempert

For more information on this publication, visit www.rand.org/t/RR1144

Library of Congress Cataloging-in-Publication Data is available for this publication.
ISBN: 978-0-8330-9659-3

Published by the RAND Corporation, Santa Monica, Calif.
© Copyright 2016 RAND Corporation
RAND® is a registered trademark.

Cover: Roberto A. Sanchez/Getty Images

Support RAND
Make a tax-deductible charitable contribution at
www.rand.org/giving/contribute

www.rand.org

Preface

This report considers how analysis can inform and support decisionmaking in complex urban areas and focuses specifically on the difficult choices among strategies and allocation of resources in response to climate change and in pursuit of greenhouse gas emissions reductions. This effort has been motivated by the analytical needs of projects that RAND is conducting in three urban areas in the United States with support from the John D. and Catherine T. MacArthur Foundation. These field studies are being used to test the practicality and utility of the methods and structuring of indicators proposed in this report. Following completion of the field studies, we will publish our results as a separate document. Our intent is for the methods and study results, if proven to be practical and useful, to be broadly applicable to early-adopting practitioners and planners interested in shaping and implementing effective urban responses to climate change.

The authors come to this topic through RAND's experience with indicators of adaptation paired with adaptation planning and decisionmaking processes in the Colorado River Basin, Louisiana, and California. Publications of interest include the following:

- David G. Groves, Jordan R. Fischbach, Nidhi Kalra, Edmundo Molina-Perez, David Yates, David Purkey, Amanda Fencl, Vishal K. Mehta, Ben Wright, and Grantley Pyke, *Developing Robust Strategies for Climate Change and Other Risks: A Water Utility Framework*, Santa Monica, Calif.: RAND Corporation, RR-977-WRF, 2014 (http://www.rand.org/pubs/research_reports/RR977.html)
- David G. Groves, Jordan R. Fischbach, Debra Knopman, David R. Johnson, and Kate Giglio, *Strengthening Coastal Planning: How Coastal Regions Could Benefit from Louisiana's Planning and Analysis Framework*, Santa Monica, Calif.: RAND Corporation, RR-437-RC, 2014 (http://www.rand.org/pubs/research_reports/RR437.html)
- David G. Groves, Jordan R. Fischbach, Evan Bloom, Debra Knopman, and Ryan Keefe, *Adapting to a Changing Colorado River: Making Future Water Deliveries More Reliable Through Robust Management Strategies*, Santa Monica, Calif.:

RAND Corporation, RR-242-BOR, 2013 (http://www.rand.org/pubs/research_reports/RR242.html)

- Alonzo L. Plough, Jonathan E. Fielding, Anita Chandra, Malcolm Williams, David Eisenman, Kenneth B. Wells, Grace Y. Law, Stella Fogleman, and Aizita Magaña, "Building Community Disaster Resilience: Perspectives from a Large Urban County Department of Public Health," *American Journal of Public Health*, Vol. 103, No. 7, July 2013, pp. 1190–1197 (http://ajph.aphapublications.org/doi/abs/10.2105/AJPH.2013.301268).

RAND Infrastructure Resilience and Environmental Policy

The research reported here was conducted in the RAND Infrastructure Resilience and Environmental Policy program, which performs analyses on urbanization and other stresses. This includes research on infrastructure development; infrastructure financing; energy policy; urban planning and the role of public–private partnerships; transportation policy; climate response, mitigation, and adaptation; environmental sustainability; and water resource management and coastal protection. Program research is supported by government agencies, foundations, and the private sector.

This program is part of RAND Justice, Infrastructure, and Environment, a division of the RAND Corporation dedicated to improving policy- and decisionmaking in a wide range of policy domains, including civil and criminal justice, infrastructure protection and homeland security, transportation and energy policy, and environmental and natural resource policy.

Questions or comments about this report should be sent to the project leaders, Debra Knopman (Debra_Knopman@rand.org) and Robert J. Lempert (Robert_Lempert@rand.org). For more information about RAND Infrastructure Resilience and Environmental Policy, see http://www.rand.org/jie/irep or contact the director at irep@rand.org.

Contents

Figures and Tables

Figures

Tables

Summary

Climate change is under way at a rate unprecedented in recorded history. Because more than 80 percent of the U.S. population lives in urban areas, cities lie at the epicenter of our nation's response to climate change. But the best strategies for U.S. urban areas to pursue, and the choice of indicators to monitor and evaluate these strategies, remain poorly understood. Developing indicators for successful urban action has proven particularly challenging because climate change is intertwined with such a wide range of a city's activities; because today's actions may have important longer-term consequences; because significant uncertainty clouds our understanding of both the scale of the impacts and the effectiveness of many responses; and because successful strategies will likely include both "low-hanging fruit" and fundamental, transformative change.

This report focuses on the front end of participatory, deliberative processes that could support decisionmaking on strategies and allocation of resources to reduce the impacts of climate change, as well as reducing greenhouse gas (GHG) emissions to limit its magnitude. For example, jurisdictions in urban areas face the following kinds of decisions.

- What types of initiatives and operational changes provide cost-effective responses to a changing climate?
- What portion of their budgets should these urban areas devote to climate-related responses, whether adaptation or mitigation, relative to other pressing public needs?
- Given a total budget for climate-related efforts, how should it be allocated among specific initiatives?
- How should various parties share the costs of adaptation and mitigation efforts?
- How should mitigation and adaptation initiatives be adjusted—for example, augmented, accelerated, or terminated—as their implementation proceeds?

To be successful, these deliberative processes require a suite of credible and publicly acceptable indicators that adequately represent the decisionmaking context, demand for services, and economic and social conditions, as well as the particulars of the external physical and biological environment. In this study, we propose a decisionmaking framework and associated indicators for urban responses to climate change based on principles of risk governance—an extension of the established practices of risk assessment and management that include considerations of governance and institutional arrangements and their economic and social context. The framework and indicators emphasize the explicit identification of capacities and processes to implement, adapt, and transform policies, institutions, financing, and other actions to effect change. We hypothesize that the framework and indicators are well-suited for characterizing how urban areas, which typically have multiple and overlapping jurisdictions, make investment and policy decisions of major consequence. These decisions are typically situated within a context of conflicting social objectives and deep uncertainty about the future.

Framework for Decisionmaking

Climate change presents a particularly difficult challenge of transformative collective action. Both limiting the magnitude and adapting to the unavoidable impacts of climate change will ultimately involve some level of transformation. Transformation generally requires significant changes by many different actors within a complex system whose future behavior is often difficult to predict with any confidence. An urban area that reduces its GHG emissions to zero or significantly alters its land use patterns to manage flooding would have had to undergo such a transformation.

We posit that risk governance, a more general form of risk management, is an appropriate decision framework with which to address such challenges because of its capacity to encompass a multiple-actor, decision-centric perspective and its recognition that climate change responses are more than technical in nature. To implement risk governance for the challenging conditions facing urban climate risk management, we add to the framework three important enhancements: decision support under conditions of deep uncertainty, iterative risk management and learning, and the idea of "tiers of transformation" to capture the varying degrees of transformation in governance that may be needed to reduce risk to acceptable levels.

Figure S.1 distinguishes among three tiers of processes that can affect the ability of urban areas to pursue climate risk management, from business-as-usual operations in the lowest tier (Tier 1) to the most challenging transformation of governance arrangements through new laws, policies, or programs in the top tier (Tier 3). Tier 1 represents actions that can be taken by existing departments in city government, existing groups, and collaborations among them, without changes in laws or any significant change in policy or budgetary allocations. Tier 2 represents actions that involve relatively minor changes in law or institutional structure at the local, state, or national level. Tier 3 involves more significant changes in laws, regulations, funding, and institutions at the state and national levels in which the cities operate; the economic environment; and

Figure S.1
Tiers of Transformation

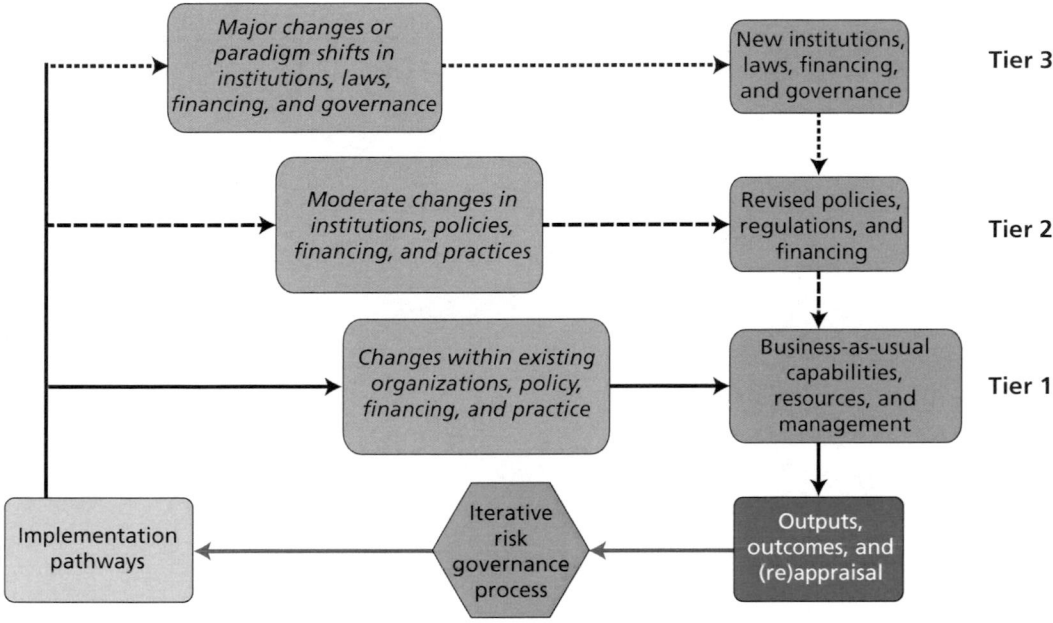

SOURCE: This figure draws on the "learning loop" framework (Kolb et al., 1975; Argyris and Schöen, 1978; Hargrove, 2002; Keen et al., 2005; Peschl, 2007; Pelling et al., 2008) often used in the resilience and climate change adaptation literature to conceptualize processes of learning and change.
RAND *RR1144-S.1*

the broader social and political context, including, for instance, trust in government, the authority of science, and the degree of civic involvement.

Climate risk management strategies that rest on actions in Tier 1 are preferable from an operational perspective to those in the higher tiers because specific local actors are often already empowered to take action, and those actions are often easier to implement. But actions in Tier 1 tend to be more limited in scope and impact than those in the upper two tiers. Change processes in these tiers pose several challenges, which are related to the concept of transformation. Changes in Tiers 2 and 3 generally involve coordinated efforts by many actors and, therefore, are harder to organize. Decision support methods and tools can help decisionmakers understand, design, and implement transformative strategies.

Indicators Within a Decision Support Process

To show how indicators could be integrated into a decision support and generalized risk governance process, we propose five general types of indicators for urban climate risk management, shown in Table S.1. These indicators are divided into two overall categories, those relating to the state of the urban region and those related to the quality of the plan and the process that produced it. The process of choosing specific indicators in the context of urban responses to climate change must always be grounded in the particular values, goals, and ambitions of the region itself. For this reason, we have focused on categories of indicators from which more specific measures could be selected.

Figure S.2, an enhanced version of Figure S.1, shows indicator categories related to the state of the urban region in green boxes and those related to the quality of the plan and the planning process in blue. The effect of current actions on changes in the tiers, and the consequences of those changes, can be addressed with various types of models.

Well-Being and Risks to Well-Being Indicators

Well-being and risks to well-being refer to conditions and risks experienced by people in the urban area now and in the future. Current conditions—such as income, health, access to recreation, and biological diversity—and some current risks—such as tidal flooding—can often be measured directly. Indicators can be drawn from such measurements. In general, however, urban climate policies have important impacts in the near and far future. Such future conditions and risks can only be estimated using some type of mathematical model. These models are inherently error-prone because of the complexities of the underlying systems and the difficulty of translating that complexity into mathematical statements. However, they have the virtue of imposing a structured

Table S.1
Proposed Categories of Indicators

Indicator Categories	Key Attributes
State of the Urban Region (Based on Causal Models)	
Well-being and risk to well-being	• Objective and subjective measures of real and perceived conditions among residents and the community at large
Capacity to implement the plan	• Management capacity • Technical skills
Capacity to adapt and transform	• Political will • Public acceptance and support • Adaptable governance mechanisms • Sustainable financial resources
Quality of the Planning	
Quality of the planning process	• Inclusivity • Legitimacy • Analytical grounding
Quality of the resulting plan	• Well-structured • Designed for evaluation and learning

Figure S.2
Enhanced Iterative Risk Management Process with Indicators

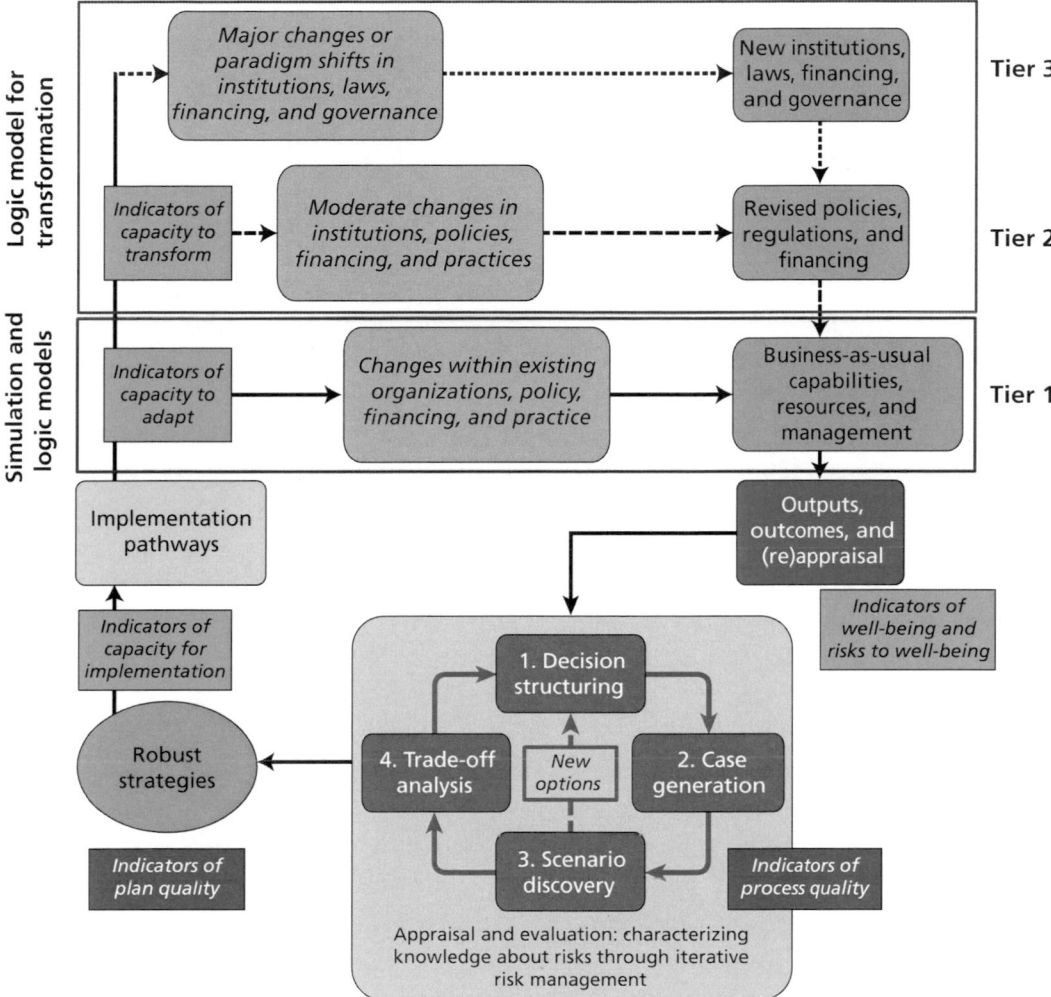

RAND *RR1144-S.2*

and transparent approach to the estimation process, using the best available science. Indicators can then be drawn from the outputs of such models.

Capacity to Implement Indicators

The *capacity to implement* the plan refers to the current resources and context available to the planners. Temporally, indicators of the capacity to implement refer to today's ability to carry out the plan and to generate outcomes that might be observed in the near term. This is a critical determinant of successful policy and includes such issues as availability of financing and funding; technical and administrative capacity within the region to manage complex design, construction, and maintenance of plan components; and sustainment of public support for the plan's implementation.

Capacity to Adapt and Transform Indicators

The *capacity to adapt and transform* refers to the urban region's ability to influence its future conditions and capabilities. We use this term to refer to both a capacity to address the impacts of climate change as they unfold and a capacity to successfully adjust policies that reduce GHG emissions as challenges and opportunities arise in the future. An urban region's capacity to adapt is limited by many fundamental attributes of the system, including its institutions, laws, human capital, and culture. The capacity to transform represents the ability to relax these constraints and open up opportunities by catalyzing significant changes in such attributes. The capacity to implement represents the ability to carry out the near-term steps of an adaptive management strategy. The capacity to adapt represents the ability to monitor and adjust course as needed in the future. The capacity to transform represents the ability to alter the institutional, legal, and other constraints on the future capacity to implement and to adapt. Many factors affect the capacities to adapt and transform, some of which may also be important in assessing the capacity to implement.

Quality of Plan and Planning Process Indicators

No set of simulation and logic models will ever provide a complete and infallible representation of the relevant urban system. In addition, the literature on decision support makes clear that the process by which information is provided can be at least as important as the information itself in improving (or not improving) decisionmaking. For instance, the extent to which information is regarded as salient, credible, and legitimate may depend on how that information is generated and provided to stakeholders and decisionmakers. Thus, it is important to complement the objective and subjective well-being indicators with indicators related to the quality of the planning process and its products.

Plan quality refers to the attributes of the plan itself. Indicators of the quality of the plan and the processes that produce it are important because they are more closely tied to factors under the planners' control than are the outcomes produced by the plan. In addition, many of the most important outcomes may be years in the future and can only be observed long after any evaluation of the plan would have occurred.

Process quality includes events leading up to the present time in a planning process. These process indicators could consist of checklists of important attributes of plans and planning processes, which, particularly for the processes, would be customized as necessary for any particular application. A planning process that included meaningful engagement with a full range of stakeholders and decisionmakers would tend to rank highly for process quality.

Example Application of the Proposed Indicator System

Given the importance of context, we consider the application of the proposed indicator system within a specific, hypothetical case of a coastal metropolis faced with major water supply, drainage, and wastewater infrastructure and land use decisions. These decisions are driven by economic development goals but are also contingent on the current and projected tax base of the region, as well as external factors, including the rate of sea level rise and risk of damage from major coastal storms and flooding.

The example urban area includes several counties and cities, each with their own land use and transportation plans. Several federal and state agencies have large equities in the region. No single regional authority coordinates planning and implementation of major public infrastructure, with the exception of a regional water management agency that has lead responsibility for some of the region's water infrastructure. Transportation, land use, and water management decisions in one county can have collateral impacts on drainage, flooding, groundwater levels, and saltwater intrusion in another. The effectiveness of changes in land use policy in one county may depend on the extent of coordination among jurisdictions within each of the counties. Decisionmakers and stakeholders are interested in understanding how development strategies and implementation of individual projects will play out under a wide range of possible scenarios that account for uncertainties in sea level rise, precipitation patterns, and land use changes.

Table S.2 summarizes the categories of indicators that might be used for this example and data sources that might enable their use. Indicator types that are new to the climate risk management literature are shown in bold. The report provides examples of indicators in the context of this particular example.

Steps Toward Applying the Framework

The first step toward implementing the risk governance framework is to find an appropriate forum for deliberative decisionmaking, or, if none exists, to create one. Forming a new governance structure for regional decisionmaking is difficult in the context of federalism. The challenges of climate change may require that such changes become more frequent when existing governance arrangements fail to demonstrate the capacity to reduce risks that transcend jurisdictional boundaries in socially and economically acceptable ways.

A second step toward implementation, closely tied to the first, is finding visionary leaders who can steer transparent and deliberative processes toward a successful conclusion by embracing democratic processes within the context of a "deliberation with analysis" approach. That is, strong leadership can set the tone for elevating objective analysis to its proper place within participatory planning and decisionmaking—not as

Table S.2
Categories of Indicators Applicable to Example

	Examples of Classes of Indicators	Approaches to Data Gathering
Well-Being and Risk to Well-Being	• Regional economic development • Flood risk • Water quality • Ecosystem protection • Financial solvency of public utilities • Equity of benefits and costs	• Environmental monitoring and socioeconomic data for current conditions • **Projections of future risk from risk assessment simulations**
Capacity to Implement (Within Tier 1)	• Financing/funding available • Incentives and their alignment for key actors • Institution, managerial, and technical capabilities • Organizational leadership and culture • Monitoring and evaluation • Public support	• Economic assessment of incentives and adequacy of financial resources • Surveys of institutional culture and public support • Qualitative assessments of other factors • **Projections of future risk from risk assessment simulations**
Capacity to Adapt and Transform	• Incentives • Public support • Leadership • **Assessment of type and magnitude of Tiers 2 and 3 changes needed to achieve benefits (e.g., risk reduction, economic growth, water quality improvement)** • **Complexity of changes**	• Economic assessment of incentives and adequacy of financial resources • Surveys of institutional culture and public support • Qualitative assessments of other factors • **Vulnerability analysis contingent on changes in Tiers 2 and 3**
Quality of Plan	• Breadth of futures and options • Use of best available science • Analysis of trade-offs among goals • Alignment of finances/funding with plan implementation • **Extent to which plan is designed for learning and adaptation**	• Checklist of plan attributes • Stock-taking of resources • **Gap analysis comparing plan to risk analysis, especially key factors driving vulnerability and potential responses**
Quality of Planning	• Openness and transparency of public process • Technical credibility • Political legitimacy • Salience to area's perceived problems and challenges • Cost-effectiveness	• Surveys and focus groups with stakeholders and decisionmakers • Tally of the resources required for the process, compared with estimated resources for other approaches with similar quality and outputs

a means to justify a predetermined decision but, rather, as a means of opening the way to consideration of a range of alternatives and trade-offs among them in terms that are clear and meaningful to the public and their representatives.

Finally, a third critical step is the harnessing of both community thought leaders and technical expertise to play leading roles in the deliberative process from the very start. Effective risk communications should begin at the very earliest stages in a planning process. Public support in turn lends credibility to the planning and decisionmaking processes. Locally based technical leadership adds further connective tissue between the public and the process. Scientists and engineers from local and regional institutions tend to have a strong understanding of local conditions and a heightened appreciation of the interplay of science and social issues within their areas by virtue of being part of the affected communities.

Measuring Effectiveness of Processes and Outcomes

In our proposed system of indicators, indicators of well-being represent the top-level outcomes that policies and strategies are intended to achieve. Practically speaking, changes in these indicators may take years or even longer to materialize as planning and implementation of response measures take shape and become reality. Estimates of the future consequences of current actions using some combination of simulation models and qualitative judgments are unavoidably and often irreducibly uncertain.

Collecting the data to populate the indicators across the five major categories listed in Table S.1 will require an evaluator who largely sits outside of the deliberative process itself to administer surveys, analyze results, and direct the collection of data and information from others sources. To be effective in measuring change—and progress—over time, an evaluator ideally is present from the creation of the process to capture baseline attitudes of participants in the planning process and identify data needs along the way.

In contrast with indicators devised primarily as "report cards" or to support a "one-off" development project, we advocate for a system of indicators embedded in long-term planning and decisionmaking processes as a means of ensuring continuity and focus. Indicators of progress are essential drivers of the processes themselves, giving resource managers and their agencies a strong motivation for maintaining the systems necessary to collect, process, and actively use the indicators with decisionmakers and the public.

Moving from Theory to Practice

Many systems of indicators have been devised for use by cities; relatively few are maintained and kept timely and fresh beyond their first few years. Even fewer systems of indicators have been tested under varying conditions and evaluated for their effectiveness. In the course of conducting this review, we have come to appreciate the need for a generalizable set of indicators that urban areas new to the long-term, integrated planning process could adapt for their own purposes. To reach that goal, however, requires further development, field testing, and evaluation.

RAND has launched three such field studies in Pittsburgh, southeast Florida, and the Bay Area megaregion of northern California to test the efficacy of the proposed approach to decisionmaking and structuring of indicators. The engagement with each area is following a "deliberation with analysis" process. Indicators are being collected to support evaluation of the planning process, the tracking of outcomes, and assessment of capacities for transformation and implementation. This measurement system will be useful for evaluation of the studies and will also enable iterative learning within the decision support processes.

Acknowledgments

The authors wish to thank Craig Howard, Director of Community and Economic Development at the MacArthur Foundation, for his guidance and support for this and ongoing work. We also would like to express our gratitude to RAND colleagues Nidhi Kalra, Melissa Finucane, Jordan Fischbach, Anita Chandra, Amy McGranahan, Kristin van Abel, and Laurie Rennie for their assistance on this study, as well as Amanda Edelman and Lauren Kendrick, fellows of the Pardee RAND Graduate School. As always, this manuscript benefited from the critique and suggestions of our reviewers, including RAND colleagues Frank Camm and Tom LaTourrette, former RAND colleague Keith Crane, and Louise Bedsworth. We wish to thank outside reviewers for their constructive comments as well.

Abbreviations

AAA	awareness, analysis, and action
CRM	climate risk management
Defra	Department for Environment, Food, and Rural Affairs (United Kingdom)
DfID	Department for International Development (United Kingdom)
GHG	greenhouse gas
HVRI	Hazards and Vulnerability Research Institute
ICLEI	International Council for Local Government Initiatives
IPCC	Intergovernmental Panel on Climate Change
MCDA	multicriteria decision analysis
MCPA	Mayors Climate Protection Agreement
MPO	metropolitan planning organization
NC	National Communications
NGO	nongovernmental organization
NI188	National Indicator 188
RBM	results-based management
RDM	Robust Decision Making
RPO	regional planning organization
SDC	Swiss Agency for Development and Cooperation
TAMD	Tracking Adaptation and Measuring Development

UK   United Kingdom

UNEP   United Nations Environment Programme

UNFCCC   United Nations Framework Convention on Climate Change

USCSP   United States Country Studies Program

CHAPTER ONE

Introduction

Cities and their surrounding areas serve as engines of regional and national economic growth, but they also amplify climate-related risk by virtue of their population density, concentration of critical infrastructure, and other high-value economic assets. In coastal cities, sea level rise and potentially increasing storm intensity pose serious threats to residents, transportation, water, housing, energy, and other infrastructure. In arid regions, extreme heat and extended drought impose stress on water supplies and riverine ecosystems. Rust-belt cities face difficult choices in revamping their aging storm and wastewater infrastructure to meet water quality requirements under changing patterns of precipitation and highly constrained public financing. Some cities are responding to these risks with efforts to adapt to a changing climate under the banner of increasing resilience to the threats of rising sea level, coastal storms, extreme precipitation events, flooding, or persistent drought (Georgetown Climate Center, 2016).

While urban areas are incurring the consequences of climate change, they also share in responsibility for the increasing threat, accounting for 54 percent of the world's population but around 70 percent of total global greenhouse gas (GHG) emissions (UN, 2014; Seto et al., 2014). The State of California and many cities elsewhere in the United States have committed to aggressive GHG emission reduction goals (State of California, 2006, 2008, 2015), consistent with the Obama administration's pledge to reduce U.S. emissions by 26 to 28 percent below 2005 levels by 2025 (White House, 2015). The promise of the Paris Agreement gives further support to these many local and state efforts (Non-State Actor Zone for Climate Action, 2016). Among many challenges, one question is how these commitments to reduce emissions will be fulfilled when these same places are wrestling with nearer-term infrastructure and other investment decisions, whether undertaken in the name of restoring aging infrastructure or improving resilience. Indeed, the line between adaptation and mitigation begins to blur as these investment decisions themselves are recognized as having long-term implications for emissions reductions, particularly those related to transportation and land use. For example, by executive order, California is addressing the state's adaptation and mitigation policies in an effort to harmonize climate goals for 2030 and 2050 (State of California, 2015).

Challenges of Decisionmaking in Response to Climate Change

The real and projected consequences of a changing climate are forcing decisionmakers to consider what actions they should take and when, how to allocate resources to support implementation, and how to share the cost burden among neighboring jurisdictions. For example, decisionmakers in urban areas are already wrestling with how to answer questions such as the following:

- What types of initiatives and operational changes provide cost-effective responses to a changing climate?
- What portion of their budgets should urban areas devote to climate-related responses, whether adaptation or mitigation, relative to other pressing public needs?
- Given a total budget for climate-related efforts, how should it be allocated among specific initiatives?
- How should various parties share the costs of adaptation and mitigation efforts?
- How should mitigation and adaptation initiatives be adjusted—for example, augmented, accelerated, or terminated—as their implementation proceeds?

Several broad challenges affect the ability of urban areas to plan, design, implement, and evaluate responses to climate change. First, there is a mismatch between established levels and structures of governance and the multiple sectors and geographic scales most appropriate to addressing climate-related challenges. As an example, southeast Florida's vulnerabilities to sea-level rise transcend the boundaries of cities and counties, which has led four counties to form a "climate compact" as a forum in which to coordinate their responses (Southeast Florida Regional Climate Change Compact, 2015). Second, there is a persistent tension between a region's short-term needs, including bouncing back quickly from natural or man-made shocks and stresses (one form of resilience), and a longer-term need to transform governance, infrastructure, and other critical urban services to reduce emissions and vulnerabilities to shocks and stresses. One of the most recent examples of this tension was felt by local officials in New York and New Jersey who sought to commit to immediate reconstruction of ocean-front boardwalks and other infrastructure in the wake of Hurricane Sandy, while the U.S. Department of Housing and Urban Development (HUD), with support from the Rockefeller Foundation, launched an ambitious "Rebuild by Design" competition to expand the options for rebuilding more resilient coastal communities (Rebuild by Design, 2016). Finally, even if the first two challenges can be overcome, there is a latent tension as governments and other organizations begin institutionalizing their responses to climate change in the near term, while recognizing their need to remain flexible and adaptive to address challenges whose scale, scope, and most effective solutions are as yet unknown.

These challenges play out in several ways. Hundreds of billions of dollars of public and private resources will be expended on adaptation and mitigation activities over the coming decades. Protection and resilience strategies alone will require costly, sustained, long-term investments with considerable uncertainty about future benefits, and their efficacy will be difficult to assess. For example, the State of Louisiana has committed to a $50 billion, 50-year master plan for coastal protection and restoration (Coastal Protection and Restoration Authority, 2012; Groves et al., 2014). As cities and surrounding jurisdictions consider actions to reduce their GHG emissions and their vulnerability to future impacts of climate change, they are obliged to quantify prospective benefits and costs of action (and inaction), as Louisiana has done. They will need to sort out how much to invest, when to invest, and when to revisit past decisions. They will need to communicate to their constituencies whether proposed actions are likely to be cost-effective and are consistent—or worthy of trading off—with other important public goals, like economic growth, equity, and environmental protection.

These can be challenging analyses to conduct, even for sophisticated public agencies in large cities, when constituent interests are often in conflict with one another and pressing near-term demands subordinate longer-term needs. Compounding the challenge, analyses of large-scale infrastructure investments are anything but ordinary because of the novelty and uncertainty of climate change, the wide range of potential responses, the long time scales that often separate actions from their consequences, and the need for actions that cross jurisdictional boundaries.

In addition to economic analyses, a consensus has yet to emerge among academics and practitioners about how to translate generally accepted notions of adaptation and resilience into actual planning guidance, design principles, and building standards. Nor has a consensus emerged about indicators of successful adaptation and resilience outcomes, although a growing literature, driven mainly by international development organizations, has sought to identify appropriate monitoring and evaluation indicators in the context of development projects (Leagnavar et al., 2015). Decisionmakers are thus hampered by lack of adequate indicators to inform planning, guide implementation, and assess the adequacy of capabilities and actual performance—in governance, technical competence, and finance—to sustain progress.

Embedding the concepts of adaptation and resilience into regional decisionmaking processes and, thus, extending beyond the usual stovepipes of urban governance is difficult. Policymakers are challenged to address needs across public service sectors and multiple jurisdictions and consider resilience as one of several critical goals of public policy. To move to this higher plane of integration, policymakers will need help in incorporating more complex forms of analysis in deliberative public decisionmaking processes and indicators to shape and guide these processes.

Indicators to Support Climate-Relevant Decisionmaking

In this study, we have drawn on current practice and literature to identify indicators that support decisionmaking processes in response to climate change in America's urban areas. We distinguish our approach from others as follows:

1. We place indicators within a specific, well-defined decision support process. This focus on decisionmaking context is often missing from other work. Like other information products that are intended to inform decisions, indicators can prove most useful as part of explicit decision processes that provide a structure and rationale for choosing the criteria. We therefore looked for forward-looking (as opposed to retrospective evaluation) indicators for implementing a decision support framework. We found that systems of indicators have been developed for many purposes (cities, development agencies, national plans) and include both process and outcome-based indicators. Some, but not all, of these systems of indicators are embedded in decision support frameworks, but most view adaptation as a linear process, rather than one with feedback loops and learning. Few indicator systems have been rigorously tested in the field. In sum, none of these existing systems is functionally suited to supporting climate-sensitive decisions in urban settings, but some have components that can be integrated into such a framework.

2. We view indicators as appropriately customized for each urban area to accommodate the heterogeneity of goals, decisions, governance arrangements, and socioeconomic conditions that is the norm rather than the exception. In contrast, some existing indicator systems have been developed by funding agencies that seek a common system to compare the effectiveness of actions across different regions.

3. We propose a decisionmaking framework that combines both process and outcome indicators. These include measurement of current conditions and quantification of future risk, contingent on near-term decisions. Simulating futures conditions in the urban system and representing those conditions with outcome-based indicators enables decisionmakers and stakeholders to consider responses to climate change that are forward-looking. While a mix of current and simulated future outcomes can address to some degree the challenge of evaluating urban climate actions over both near-term and longer time scales, outcome indicators are necessarily uncertain and incomplete. Thus, process indicators also are necessary to track changes that are likely to enable progress over time.

4. We seek to lift climate-related decisions out of the mitigation, adaptation, and disaster resilience boxes in which they are often placed. Instead, we treat these goals as complementary to a wider range of decisions related to the built environment and community development.

5. We focus on the capacity of urban areas to transform their governance and operations to meet the climate challenge. Many responses to climate change can be accomplished within current jurisdictional structures and with small and incremental changes to current practices. But some future scenarios are likely to require transformative responses from individuals, government, and the private sector that extend well beyond the current planning, operational, and financial capabilities of urban areas. Responding to climate change will often also require a more integrative approach to urban infrastructure planning, design, and operations that fragmented and stovepiped city departments will be hard-pressed to handle. Thus, the decisionmaking framework and indicators proposed here include a focus on an urban area's capacity for transformation and an explicit accounting of how near-term actions might influence such changes.

Overall, the proposed decision-centric approach to indicators aims to improve communications and engagement with stakeholders and enhance the ability of public agencies and other organizations to take appropriate and well-informed actions. Further, with this approach, we suggest a means of embedding responses to climate change in a larger decision context that encompasses both missions and longer-term transformations of urban areas.

Examples of Climate-Related Decisions Faced in Urban Areas

To ground our discussion of indicators in the real problems faced in urban areas, we provide three examples of cases in which they might be used. The proposed decisionmaking framework and indicators in this report aim to address the kinds of analytical needs raised by each of these examples of climate-relevant decisionmaking in urban areas. We return to these examples later in the report when illustrating the prospective application of the proposed system of indicators.[1]

Case 1: Adapting and Upgrading Aging Infrastructure

Imagine a now-thriving rust-belt city that faces aging infrastructure and significant wet weather flooding and sewage overflow problems, exacerbated by changing storm intensities. The region has a fragmented governance structure, with one large city surrounded by over a hundred smaller municipalities. Many jurisdictions favor innovative, green infrastructure solutions—a departure from the region's historical focus on large infrastructure projects. Regional leaders are developing a comprehensive, long-term plan. With uncertainties about the performance of new approaches and a chang-

---

[1]   Testing of the indicators in the field is in progress and will be reported in a subsequent publication.

ing climate, how can the region's leaders ensure themselves and their community that their plan will be effective?

Case 2: Coping with Sea-Level Rise

Imagine a heavily populated, low-lying coastal region facing rising sea levels that lead to frequent flooding, threaten major infrastructure, complicate environmental protection, and increase the risk of disastrous storm surge from major hurricanes. The region's counties and cities have agreed on goals and guidelines for a common response to sea-level rise. Individual jurisdictions are now implementing policies and are beginning to incorporate these guidelines into major flood control, water supply, and transportation infrastructure projects. The scale and timing of future climate impacts remains uncertain. How does each jurisdiction ensure that the actions it takes today are consistent with its own near- and long-term goals, as well as consistent with the actions of neighboring jurisdictions with whom it shares a common, managed coastal and hydrologic system?

Case 3: Shifting Land Use and Transportation to Reduce Emissions

Imagine a sprawling metropolitan region committed to significant reductions in GHG, much of which emanate from its transportation sector, land use patterns, and built environment. The state has required metropolitan planning authorities to develop long-term plans that ensure that land use and transportation systems are consistent with emission reduction targets. However, most of the funding for such decisions flows through the cities in the region, which generally share climate and many other goals but also compete with each other for jobs, economic development, and investments. Many aspects of transportation and land use, including the fundamental underlying technologies, the costs of emission reductions, and future funding models, remain deeply uncertain, while current infrastructure decisions can have implications that span decades. How do multiple jurisdictions within the region ensure that their plans are consistent with deep reductions in GHG emissions, while at the same time maintaining their local autonomy and prerogatives?

Definition of Frequently Used Terms

In addressing indicators for urban climate risk management, definitional issues abound. Actions to address climate change are generally divided into two broad categories of mitigation and adaptation. *Mitigation* refers to efforts to limit the potential long-term consequences of climate change by reducing the concentration of GHG in the atmosphere through emissions reductions and land use changes. *Adaptation* refers to processes of coping with such changes as sea-level rise, extended drought, and shifts in ecosystem dynamics already under way and those anticipated under a range of assumptions about atmospheric concentrations of GHG. Many countries and individual cities are pursuing both mitigation and adaptation strategies, and, in fact, the historical dichotomy between these two paths is proving less useful as some actions and strategies, like improved land use planning, can both reduce risk and reduce net GHG emissions.

Indicators, Measures, and Metrics

We have chosen to focus on indicators in this study. One common definition of *indicator* is a "measurable representation of the condition or status of operations, management or conditions" (ISO, 2015). The Organisation for Economic Co-operation and Development (OECD) defines an *indicator* as a "quantitative or qualitative factor or variable that provides a simple and reliable means to measure achievement, to reflect the changes connected to an intervention, or to help assess the performance of a development actor" (OECD, 2002). Many others have proposed similar definitions.

In contrast, the terms *measure* and *metric* have more precise quantitative meanings and are thus subsets of the more generic *indicator*. A *measure* is "a value that is quantified against a standard." The standard could be in terms of dollars, tons of emissions, acres, or other similar units (Viggh, 2015, p. 73). A *metric* is "a calculated or composite measure or quantitative indicator based upon two or more indicators or measures" (Viggh, 2015, p. 73).

In this work, we are interested in identifying and classifying both quantitative and qualitative indicators, some but perhaps not all of which may rise to the level of measures or metrics. For instance, indicators might include quantitative data on emissions levels, water quality, flood risk, and funding available for relevant programs. Indicators might also include qualitative judgments of institutional capacity and survey data on citizens' satisfaction with a planning process. As we define these terms in Table 1.1, indicators are a means of consistently tracking inputs; outputs; and short-, medium-, and longer-term outcomes using replicable and credible methods of evaluation and assessment.

Indicators to capture the long-term outcomes of mitigation actions are relatively easy to identify, although not necessarily easy to measure in practice—namely, atmospheric concentrations of GHG and their rate of change, and both global and country-specific emissions by mass or by changes in emissions intensity. These indicators are now memorialized in the form of voluntary national commitments to emissions reductions secured in the Paris climate agreement (United Nations Framework Convention on Climate Change [UNFCCC], 2015). Indicators that capture the long-term outcomes of adaptation and resilience activities are more problematic for the reasons cited above. In addition, indicators of outputs and outcomes of transformational processes in governance, law, and regulation are qualitative in nature and even more difficult to nail down. Nonetheless, these types of process-oriented indicators are likely to be similar for mitigation and adaptation decisionmaking, somewhat simplifying our task.

Adaptation, Resilience, and Transformation

Developing indicators for adaptation to climate change is challenged by a host of complexities, not the least of which is that the definitions of these terms are in flux. In

Table 1.1
Definition of Terms Relevant to Indicators

Term	Definition
Indicators	Measurable representations of inputs, outputs, and outcomes of operations, management, strategies, and the status and trends of internal and external conditions (ISO, 2015)
Inputs	Resources generally defined (e.g., money, staff time, reputation) that are used or consumed when an organization implements an action or process. Inputs affect outputs and can potentially affect outcomes (USAID, 2009).
Measures	Values that are quantified against a standard. The standard could be in terms of dollars, tons of emissions, acres, or other similar units (Viggh, 2015, p. 73).
Metrics	Calculated or composite measures or quantitative indicators based upon two or more indicators or measures (Viggh, 2015, p. 73)
Outputs	Observable products or conditions over which an organization exercising a process has direct control and that can potentially affect outcomes (OECD, 2002)
Outcomes	Attributes of the world that policymakers and decisionmakers care about, such as the health and safety of a region's residents or the preservation of biodiversity. Outcomes are generally related to outputs, but the connection may not always be direct (authors' definition).
Processes	Mechanisms by which inputs are transformed into outputs in a way that the organization exercising the process typically understands and can define (ISO, 2015)

fact, there is currently significant overlap and tension between the concepts of climate change adaptation and resilience. The Intergovernmental Panel on Climate Change (IPCC, 2014c, p. 118) definition of *adaptation* is

> the process of adjustment to actual or expected climate and its effects. In human systems, adaptation seeks to moderate or avoid harm or exploit beneficial opportunities. In some natural systems, human intervention may facilitate adjustment to expected climate and its effects.

In a similar vein, the National Research Council (NRC, 2010, p. 1) defines *adaptation* as referring to

> adjustments in ecological, social, or economic systems in response to actual or expected climatic stimuli and their effects or impacts. It refers to changes in processes, practices, and structures to moderate potential damages or to benefit from opportunities associated with climate change.

The IPCC defines *resilience* as "the ability of a social or ecological system to absorb disturbances while retaining the same basic structure and ways of functioning, the capacity of self-organization, and the capacity to adapt to stress and change" (IPCC, 2013). Tyler and Moench (2012, p. 312) argue for resilience as a better goal than adaptation per se:

The standard approach to planning for climate adaptation . . . is to frame the task as adjusting policies, practices and plans in order to avoid negative impacts of climate change. In essence, this approach relies on prediction as the basis to identify avenues for prevention. Instead of focusing on discrete measures to adapt to specific perceived future climate risks, it may be more effective for cities to consider the problem as one of building resilience. In the case of urban climate adaptation, an approach based on resilience encourages practitioners to consider innovation and change to aid recovery from stresses and shocks that may or may not be predictable.

Yet resilience has also come under criticism with the view that its use often implies that the current socioeconomic paradigm will continue. Bours et al. (2013) suggests instead that responses may need to be transformative and encompass "the ability to adjust to potentially radical changes in context, not just withstand shocks" (Bours et al., p. 60). Transformational responses apply to both mitigation and adaptation.

For the purposes of our study, we will hew to the definition of terms in most common use. Table 1.2 summarizes the definitions of relevant terms listed in the Annex III Glossary from the IPCC's Fifth Assessment Report (IPCC, 2013).[2] A more complete collection of definitions can be found in Appendix A.

We find it clarifying to think of adaptation as a general descriptor of planning and implementation processes to achieve any number of goals, with resilience being one of the most important and easy to communicate of those goals to the public. Adaptive capacity in risk governance is an enabler of successful adaptation activities undertaken to reduce vulnerabilities and increase resilience. Transformation is the more radical concept, suggesting a highly consequential paradigm shift in both policy and governance.

How This Report Is Organized

In Chapter Two, we describe a risk governance decisionmaking framework and contrast it with other paradigms for decisionmaking within complex systems. Within the risk governance framework, we propose "tiers of transformation," a structured approach to thinking about how urban areas can and do transform their way of doing business to achieve better outcomes than would be achieved under a business-as-usual mode of operating. In Chapter Three, the risk governance decisionmaking framework provides our "baseline" context for evaluating the applicability of existing systems of indicators to climate-relevant decisions. We then describe how to apply the framework and indicators to a hypothetical case of a coastal metropolis attempting to tackle major climate challenges in a more regionally coherent way. Chapter Four summarizes our findings and next steps in evaluating the efficacy of our proposed approach.

---

[2]   Each of these terms has a more general meaning outside of the context of climate change. Given the focus of this review, however, we use the narrower, climate-specific definition.

Table 1.2
Definitions of Terms Relevant to Climate Risk Management

Term	Definition
Adaptation	"The process of adjustment to actual or expected climate and its effects. In human systems, adaptation seeks to moderate harm or exploit beneficial opportunities. In natural systems, human intervention may facilitate adjustment to expected climate and its effects." Incremental adaptation refers to "[a]daptation actions where the central aim is to maintain the essence and integrity of a system or process at a given scale" (Park et al., 2012). Transformational adaptation refers to "[a]daptation that changes the fundamental attributes of a system in response to climate and its effects."
Adaptive capacity	The ability of systems, institutions, humans, and other organisms to adjust to potential damage, to take advantage of opportunities, or to respond to consequences
Mitigation (of climate change)	"A human intervention to reduce the sources or enhance the sinks of greenhouse gases." Mitigation can be accomplished through technological change and substitution that reduce resource inputs and emissions per unit of output.
Resilience	The capacity of a social-ecological system to cope with a hazardous event or disturbance, responding or reorganizing in ways that maintain its essential function, identity, and structure, while also maintaining the capacity for adaptation, learning, and transformation (Arctic Council, 2013)
Risk	The potential for consequences where something of value is at stake and where the outcome is uncertain, recognizing the diversity of values. Risk is often represented as the probability of occurrence of hazardous events or trends multiplied by the impacts if these events or trends occur. Risk results from the interaction of vulnerability, exposure, and hazard. In [the IPCC] report, the term risk is used primarily to refer to the risks of climate-change impacts.
Risk management	Plans, actions, or policies to reduce the likelihood and/or consequences of risks or to respond to consequences
Transformation	Altering the fundamental attributes of a system, including value systems, regulatory, legislative, or bureaucratic regimes, financial institutions, and technological or biological systems
Urbanized area (or urban area)	According to the U.S. Census Bureau, urbanized areas comprise one or more places ("central place") and the adjacent densely settled surrounding territory ("urban fringe") that together have a minimum of 50,000 persons. The urban fringe generally consists of contiguous territory having a density of at least 1,000 persons per square mile (U.S. Census Bureau, 1995).
Vulnerability	The propensity or predisposition to be adversely affected. Vulnerability encompasses a variety of concepts including sensitivity or susceptibility to harm and lack of capacity to cope and adapt.

SOURCE: IPCC, 2013, unless otherwise noted.

NOTE: The complete set of terms is available from the Chief, Geography Division, U.S. Bureau of the Census, Washington, D.C., 20233.

CHAPTER TWO

Risk Governance Framework for Decisionmaking

Urban areas seeking to respond to climate change face a range of challenges, as evidenced by the three examples given in Chapter One. Both mitigation and adaptation responses must engage a wide array of actors, affect both near- and long-term change, and find an appropriate balance among shared benefits and outcomes. A variety of decisionmaking frameworks to guide response actions have been proposed for climate change adaptation and mitigation policies. We review such frameworks in Appendix B. In this chapter, we describe a generalized form of risk governance as a particularly useful and broad-gauged approach toward developing a robust system of indicators. We seek indicators of both means and ends. Some indicators are intended to characterize progress and success of specific decision support and decisionmaking processes associated with the choice of urban climate response strategies. Other indicators are intended to capture outcomes and outputs of the strategies once implemented.

Enhancing Risk Management

The IPCC Fifth Assessment Report found that climate change is best viewed as a challenge of risk management (IPCC, 2014b; Jones et al., 2014). Other recent assessments echo that view (NRC, 2010; IPCC, 2012; Moss et al., 2014). These reports suggest that when appropriately generalized, risk management provides an inclusive framework that enables assessment and responses to the challenges and opportunities generated by climate change.

Risk is defined as the product of probability of an event or phenomenon affecting an individual, community, population, or place and the consequence of occurrence of that event or phenomenon; a very rare event with large consequences may have similar risk as a much more frequent event with moderate consequences. The literature identifies three main contributors to risk: *hazard*, which is the potential occurrence of a physical event that may cause injury or damage; *exposure*, which is the presence of people and the things they care about in places that could be adversely affected; and *vulnerability*, which is the propensity to be adversely affected by the hazard. For instance, the potential for a flood would represent hazard, the people and built environment in the flood plain would represent exposure, and the likelihood of structural damage and inability to evacuate would represent vulnerability. Hazard, exposure, and vulnerability can each have associated probabilities. Together they lead to the consequences that constitute risk. Distinguishing among these three contributors is useful because the human consequences arising from any physical hazard are strongly mediated by social factors, such as where people live and the capabilities available to them to mitigate exposure and vulnerability. Risk management involves efficiently allocating resources among actions to reduce hazard, exposure, and vulnerability—thus offering a rich set of options for reducing risk.

Risk management is one of several frameworks. Others discussed in Appendix B include the precautionary principle, resilience, vulnerability, and control theory. Of the frameworks considered, risk management seems best for jurisdictions seeking to enhance the well-being, opportunities, and quality of life of their residents in the face of climate change and other important stressors. The risk management framework emphasizes a decision-centric view with multiple actors, which seems most applicable to the urban environment. It offers a large body of practice from which to draw methods and experience. In addition, resting on its foundation of hazard, exposure, and vulnerability, risk management can incorporate key insights from other frameworks, as discussed in Appendix B.

But risk management in practice often has narrow connotations, with numerous examples of a technocratic perspective, separated from the institutions and processes of governance and community dynamics that inform and influence choices made by decisionmakers. For this reason, we think it is vital to employ risk management within a broader framing of risk governance. In his 2008 book *Risk Governance: Coping with Uncertainty in a Complex World*, Ortwin Renn argues that risks appear in a broader context of human choices that aim to satisfy human aspirations, wants, and needs. In pursuing these ends, humans' actions create consequences, intended and unintended. The contingency of such consequences introduces uncertainty, and such uncertainty is intrinsic to the concept of risk. Consequences are often adverse, but the risk concept also includes contingent consequences that may be beneficial.

Risks arise from a mix of individual and collective human actions. Thus, managing risk involves understanding and interacting with a broad array of actors, enabled and constrained by institutional arrangements; organizational capacity; competing demands for resources; and political, social, and regulatory cultures. Ultimately, Renn writes, risk is central to our understanding of human agency—that is, our ability to act in a strategic fashion linking our decisions with outcomes, intended and unintended, likely and unlikely (Renn, 2008, p. xiii).

Risk governance focuses on the processes by which such groups as government agencies, businesses, citizens' organizations, and technical experts make risk management decisions. In the context of urban responses to climate change, the framework emphasizes three particularly important ideas. First, risk governance retains the core goal of efficient resource allocation to achieve desired outcomes in the face of uncertain future events, while placing the decisionmaking in a much broader context.[1] Second, risk governance emphasizes the social construction of risk that arises from both the physical consequences of potential events and human judgments regarding the seriousness of these events, the legitimacy of the activities and entities that may give rise to those events, and the social value of the activities that may be required to ameliorate the resulting risks. Third, risk governance takes an expansive view of the context in which groups of individuals come to recognize risks and to identify and take actions in response.

This context includes the totality of actors, rules, conventions, processes, and mechanisms relevant to how risk information is collected, analyzed, and communicated and how responses are identified, evaluated, chosen, and implemented. The context also includes institutional arrangements, such as regulatory and legal frameworks; coordination mechanisms, such as markets and norms; and political culture (Renn, 2008, p. 374).

As shown in Figure 2.1, risk governance consists of an ongoing sequential process of preassessment (initial problem scoping), appraisal (analysis of the problem and identification of alternatives), characterization/evaluation (analysis of alternatives), and management (choice and implementation) (International Risk Governance Council [IRGC], 2006). Risk communication accompanies each of these phases. While these are the standard elements of just about any decisionmaking framework, Renn describes this simple depiction of the elements of risk governance as "open, cyclical, iterative, and interlinked" (Renn, 2008, p. 47), in contrast with more common linear structures of decisionmaking. Most important, the approach implies an intertwining of science-based information and societal values by labeling the left side of the graphic as "understanding" (generating and collecting knowledge about risks) and the right side as "deciding" (making decisions about how best to manage risks).

In the preassessment phase, the problem is framed and defined, a process that necessarily involves societal values as well as the best available information regarding the risks and responses. The appraisal and characterization/evaluation phases—which entail understanding the extent to which certain risks are or are not tolerable in the present and the trade-offs among risks and potential responses to reduce risks in the future—similarly intertwine both quantitative information and societal values. Finally, the management phase draws on the analysis of alternatives in the previous step to enable decisionmakers to choose among alternatives in a way that strikes an acceptable balance of risk and reward among the multiple social goals.

---

[1] Expressed in the context of multicriteria decision analysis, efficient resource allocation in the economic sense is one of many objectives that society might pursue (e.g., environmental quality, equity, regional economic development), seeking through a legitimate process of social choice to achieve an appropriate balance among these objectives (Sen, 2009).

Figure 2.1
Risk Governance

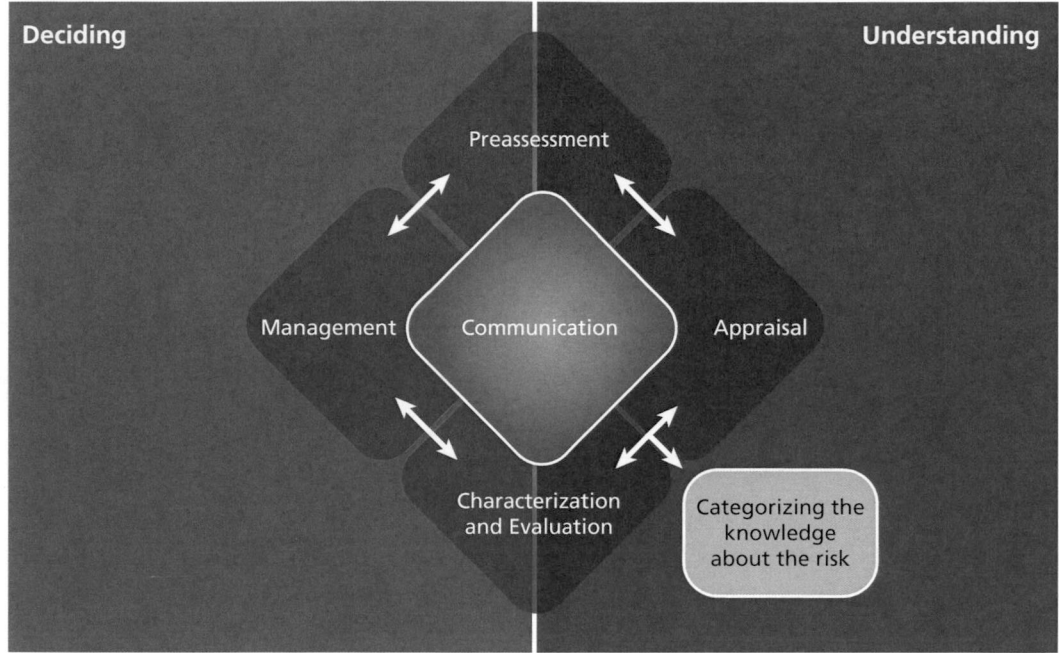

SOURCE: IRGC, 2006.
RAND *RR1144-2.1*

Overall, risk governance encompasses the way in which societies make collective decisions about technologies and activities that have uncertain consequences. The approach is particularly germane in situations in which there is no single authority able to take binding actions, so that managing risk requires cooperation and collaboration among many actors. Such situations are ubiquitous in urban responses to climate change, particularly so in democratic societies. Indeed, context is critical, as illustrated in Figure 2.2 through the ever-larger circle of players and cultural features that influence decisionmaking.

Appendix B discusses complementary approaches to decisionmaking or, in some cases, alternatives to the risk governance framework. We highlight concepts that are often discussed in the context of responses to climate change, like vulnerability and resilience, as well as other concepts, such as control theory and game theory, that are less commonly invoked but no less useful in thinking about complex urban systems.

Figure 2.2
Risk Governance in Context

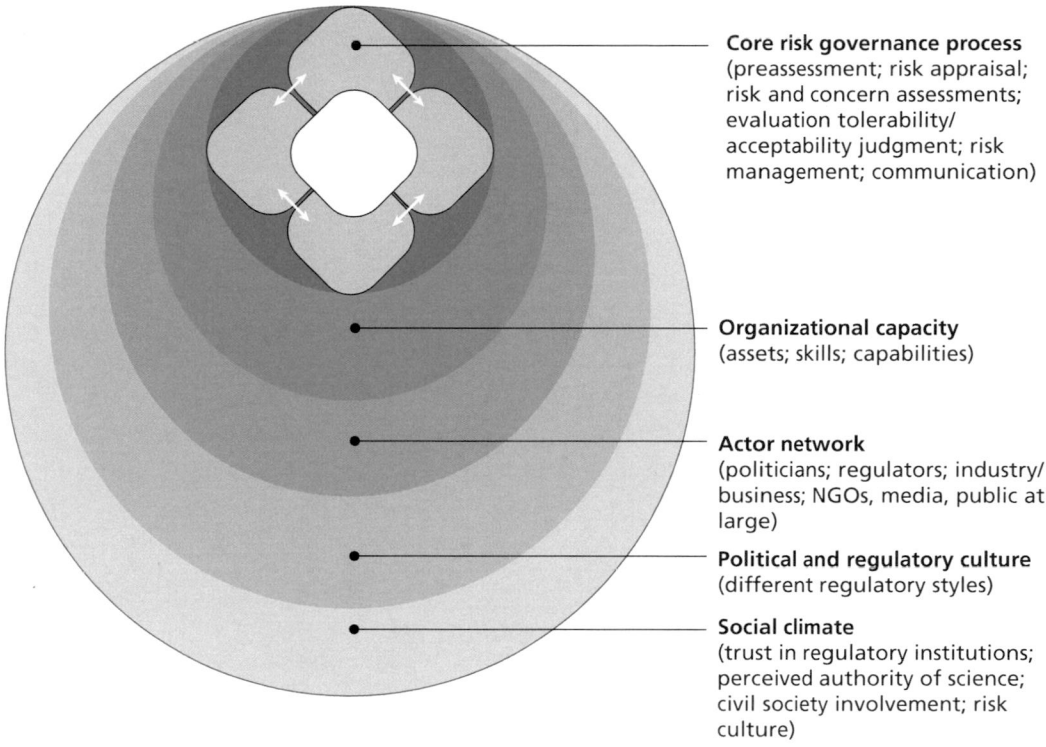

Core risk governance process
(preassessment; risk appraisal;
risk and concern assessments;
evaluation tolerability/
acceptability judgment; risk
management; communication)

Organizational capacity
(assets; skills; capabilities)

Actor network
(politicians; regulators; industry/
business; NGOs, media, public at
large)

Political and regulatory culture
(different regulatory styles)

Social climate
(trust in regulatory institutions;
perceived authority of science;
civil society involvement; risk
culture)

SOURCE: IRGC, 2006.
NOTE: NGO = nongovernmental organization.
RAND RR1144-2.2

Enhancing Risk Governance

Implementing a risk governance framework in real-time decisionmaking contexts requires practical analytical tools and political acumen. In this section, we highlight three important enhancements of risk governance in an urban context: decision support under conditions of deep uncertainty, iterative risk management and learning, and the potential need for transformational strategies.

Effective Decision Support Under Conditions of Deep Uncertainty

Effective urban risk governance will generally require appropriate use of quantitative information. Without quantitative information, identifying and assessing risks and efficiently and effectively allocating resources to reduce them can be difficult. Data to inform such choices have become increasingly available; for instance, information on current and future climate has proliferated, and cities are learning how to exploit "big data." Literature and practice make abundantly clear, however, that good scientific and technical data alone are insufficient for good decisionmaking (Renn, 2008; Jones et al., 2014). In particular, in the rich decision contexts of urban risk governance, significant tensions often exist between use of such information and the way decisions are actually made. Quantitative estimates of risk by experts may differ from those of lay people, in some cases because they provide new information, but in others because they neglect attributes important to the public. Quantitative information may not be relevant, timely, or understandable for the choices being made. Numerous examples exist of decisions that might have been improved had those making them had better access to information at the time the decision was made. In one particularly tragic example, virtually all of the emergency backup generators in New Orleans were located in the basements of hospitals in New Orleans, even though the city's vulnerability to storm surge and coastal flooding was well known to technical experts (and many older residents) at the time Hurricane Katrina struck in August 2005.

The concept of decision support (Moss et al., 2014) is useful for understanding how quantitative (and other) information can most effectively contribute to risk governance. The National Research Council (2009) defines decision support as a "set of processes intended to create the conditions for the production and appropriate use of decision-relevant information." The science of decision support aggregates understanding from organizational behavior, psychology of judgment and decisionmaking, and evaluations of a large body of practice in many policy areas. One review of this literature (NRC, 2009) suggests key principles of effective decision support particularly relevant to urban climate programs. These include the following:

1. the priority of process over products—that is, the way in which quantitative information is integrated into decisionmaking processes often is as important as the information products themselves
2. the need to link information producers and users
3. institutional stability in the provision of information
4. the need to design for learning.

In particular, these principles guide thinking about how various indicators can provide useful information to different types of decisionmakers, enabling them to make better decisions.

In risk governance, people can bring a wide variety of differing perspectives to the identification, evaluation, and management of risks. In some cases, quantitative information can significantly narrow these differing perspectives. But in many situations, the quantitative information can do little to narrow such perspectives because it is too uncertain, too focused on only some aspects of the situation, too rooted in particular communities of knowledge producers to the exclusion of others, and too intertwined with diverging values. In addition, people often bring to discussions of risk management interlocking clusters of values, policy preferences, and expectations about the future. People will tend to prefer policy actions consistent with their values and, thus, tend to trust quantitative information that describes risks that can be treated with actions consistent with those values (Kahan and Braman, 2006).

Various decision support methods exist that can usefully inform urban climate risk management. Of particular interest here is the use of Robust Decision Making (RDM) within a multiattribute decision framework. Multiattribute decision theory (Keeney et al., 1993), sometimes called multicriteria decision analysis (MCDA), provides a general framework for comparing the consequences of alternative policy choices when several, often conflicting, policy objectives are in play. This is in contrast to such approaches as benefit-cost analysis that aggregate all potential consequences into a single (often monetary) indicator. Multiattribute decision theory acknowledges that people may care about different potential consequences of their actions, and different people may not find it useful to combine all such consequences into a single number. Decision support tools based on MCDA will thus retain multiple indicators and present users with trade-off curves or tables that allow them to evaluate policy options based on trade-offs among these indicators. For instance, RAND's support for Louisiana's Comprehensive Master Plan for a Sustainable Coast included a multiattribute Planning Tool that considered coastal storm risk reduction in terms of expected annual damage, loss of land expressed in acres, cost of actions to mitigate risks and land loss, and other indicators important to the public (Groves et al., 2013).

In addition to multiple objectives and differing values, stakeholders will often bring differing expectations about the future. Traditional probabilistic decision and risk analysis (Morgan et al., 1990; Jones et al., 2014; Moss et al., 2014) can sometimes fare poorly under such conditions because it requires all parties to agree on a single set of assumptions about the future before they can accept both the framing and results offered by the analysis. Under the conditions of deep uncertainty that often characterize urban climate risk management, following the reverse process may in fact be a more useful construct (Kalra et al., 2014).

RDM (Lempert et al., 2003; Lempert et al., 2006; Hallegatte et al., 2012) provides one such "reverse" approach. RDM begins with a proposed plan or plans and then uses analytics to stress-test them over thousands or millions of alternative paths into the future. RDM adopts a modeling approach that summarizes the conditions in which each plan will work well or poorly. The RDM process, represented in Figure 2.3, shares features with risk governance through its feedback processes and distinctions between processes of understanding and decisionmaking. The process begins with an interactive, participatory process with decisionmakers and stakeholders to define the nature of the decision: goals, metrics, uncertainties, potential strategies and options, and available modeling tools. Once the decision is structured for analysis, the second step is to run a model of the system of interest many times to project future conditions across the full range of uncertain factors. This provides the database for the third step of the vulnerability assessment, in which future conditions are identified when the system (either as it currently exists or with some combination of possible new options)

Figure 2.3
Steps in Robust Decision Making and Their Relationship to Steps in Risk Governance

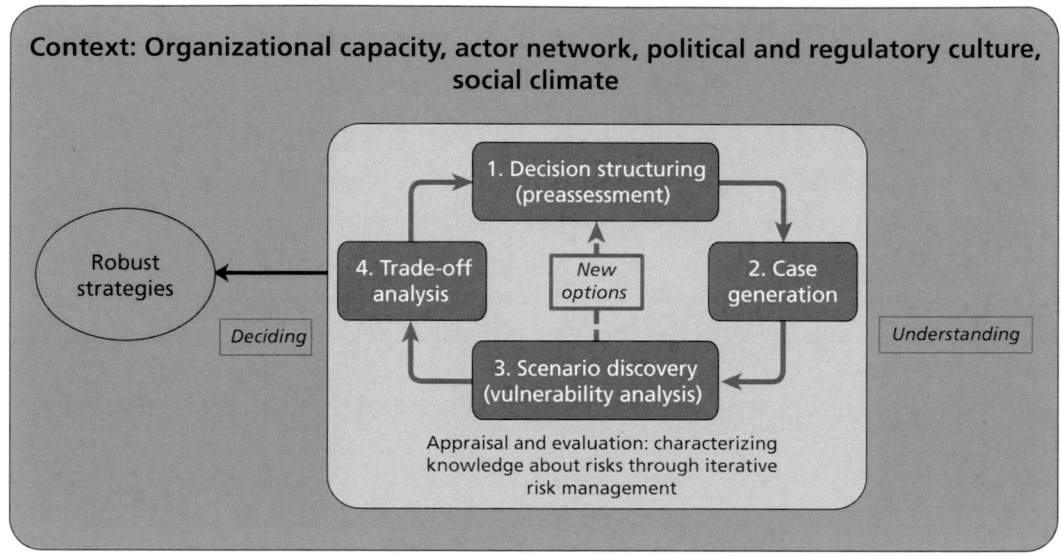

RAND *RR1144-2.3*

performs poorly. This is sometimes called "scenario discovery" and leads to identifying new options or strategies that might be able to reduce the vulnerabilities that have been identified. The fourth and last step is when the analysis advances to visualizations of trade-offs across the multiple objectives. The aim is to identify robust strategies that perform well for the objectives of interest across a wide range of possible future conditions.

RDM uses analytics to explore consequences without necessarily privileging one assumption about the future over another. RDM can significantly enhance the value of simulation models initially designed for predictive analysis by running them over many plausible paths into the future in order to identify vulnerabilities of proposed strategies and potential robust responses. The approach also provides output in a scenario-based form that helps decisionmakers agree on which plans to pursue without requiring prior agreement on assumptions. This can reduce conflicts among stakeholders as well as facilitate interagency processes (Fischbach et al., 2015).

Drawing on the discussion in Fischbach et al. (2015), RDM explicitly follows a "deliberation with analysis" process of decision support, in which parties to the decision deliberate on their objectives, options, and problem framing; analysts generate decision-relevant information using the system models; and the parties to the decision revisit their objectives, options, and problem framing influenced by this quantitative information (NRC, 2009). When used with MCDA, the RDM process aims to facilitate deliberation among diverse stakeholders by embedding systematic quantitative reasoning about the consequences of, and trade-offs among, alternative decision options within a framework that recognizes the legitimacy of different interests, values, and expectations about the future (Lempert, 2013; Parker et al., 2014).

Such processes have been used to facilitate Louisiana's Comprehensive Master Plan for a Sustainable Coast (Groves et al., 2013), the U.S. Bureau of Reclamation Colorado Basin Supply and Demand Study (Groves et al., 2012), and other engagements (RAND Water Climate and Resilience Center, 2016). The approach seems particularly appropriate for decision support for urban climate risk management. For example, this process, similarly to Renn's preassessment phase, would begin with a decision-scoping exercise with stakeholders, facilitated by the analysts. In this exercise, the parties to the decision define the objectives and indicators of the decision problem, policy options that could be used to meet these objectives, the uncertainties that could affect the success of proposed plans, and the relationships that govern how plans would perform with respect to the indicators. Importantly, this scoping step aims to clearly differentiate between factors under and outside decisionmakers' control. The former can become part of the strategy that decisionmakers pursue, while the latter include the uncertainties outside decisionmakers' control. As one important decision, the scoping exercise could identify one or more proposed urban adaptation plans (the set might include business as usual) for the region that will be considered in subsequent steps.

Determining the goals of the decision process is a nontrivial part of this scoping process. In the private sector, where profit maximization is the goal of business strategy and decisionmaking, outcome indicators for profit-making and investment returns are well-established. This is decidedly not the case for the public sector, which typically seeks to achieve multiple and often competing goals through its actions. These goals may include the provision of public safety and other basic services, improvement of economic and social well-being of all residents, some measure of equity in opportunity across groups and communities, and increasing resiliency and reducing or otherwise managing risks from disruptive events—whether short-term and discrete, like a devastating storm, or long-term and slow-moving, like climate change. In these decision-making contexts, widely accepted indicators are still evolving, to the extent that they exist at all.

The scoping exercise also helps identify simulation models and data that analysts can then use to evaluate proposed adaptation plan(s) over many plausible paths into the future. Different combinations of the uncertainties define these paths. The simulation runs generate a large database of simulation model results. Analysts and decisionmakers can then use visualizations and statistical "scenario discovery" algorithms on this database to explore the simulation data and identify the key combinations of future conditions in which each candidate plan would meet or miss decisionmakers' objectives. This approach can also provide some of the most important benefits of scenario analyses. These include helping stakeholders with differing values, interests, policy preferences, and expectations about the future to explore potentially inconvenient or unsettling futures; to imagine a wider range of plausible futures and expand the range of solutions they consider; and to potentially reach consensus on a plan even if they continue to disagree on how the future might unfold.

Using these scenarios, decisionmakers can explore the trade-offs among alternative plans. They may choose a robust strategy or decide that none of the alternatives under consideration proves sufficiently robust. In this latter case, they could return to the search for suitable candidates, perhaps through modification of or hybridization among the initial set, this time with deeper insight into the strengths and weaknesses of the alternatives initially considered.

The steps of the RDM process are designed to represent a *deliberation with analysis* process. For instance, the decision framing step involves deliberation, using the database of simulation runs to identify scenarios that illuminate vulnerabilities, and then using these scenarios to judge proposed plans and identify new ones, followed by a trade-off analysis.

Iterative Risk Management, Learning, and Adaptive Planning

Risk governance includes the concept of *iterative risk management*, which involves a process of assessment, action, reassessment, and response that will continue, in the case of many climate-related decisions, for decades, if not longer (Jones et al., 2014). In particular, urban climate policies will necessarily evolve over time in response to new information because uncertainty surrounds many important aspects of the urban climate challenge, so that future decisionmakers will find many choices made today poorly suited to conditions at that time. While uncertainties about future climate are clearly important, those regarding the evolution of each city and the effectiveness of alternative risk management actions may be even more consequential, because cities today are undergoing a process of significant economic, social, technological, and physical transformation.

Recent years have seen a growing interest in formalizing the concepts, methods, and tools for developing and evaluating adaptive decision strategies. Such strategies, which are designed to evolve over time in response to new information, are ideally those that emerge from a successful iterative risk management process (Walker et al., 2003; Lempert et al., 2010; Groves et al., 2013; Haasnoot et al., 2013). This formalization of adaptive strategies may help inform the development of urban climate risk management programs and the indicators needed to evaluate them. As one key distinction, this literature emphasizes the difference between planned and unplanned learning (NRC, 2009). In the latter, organizations respond to events as they occur but devote little attention or resources to understanding how to make the learning process more effective and durable. The climate change adaptation literature similarly distinguishes between the terms "coping" and "adapting" and between planned and unplanned adaptation (IPCC, 2014b).

Unplanned learning can be costly. It can be more expensive over the long term, can impede organizations from achieving their goals, and can increase the chance of catastrophic failures. Unplanned learning also can make it more difficult to promote consensus among parties to a decision who bring significantly different expectations about how the future may unfold. But consistent with the observations of Preston et al. (2011), cities may pursue unplanned adaptation because it has some important (if transitory) advantages. When budgets are limited and election cycles short, solving current problems is often more compelling than investing to increase the ability to change policies in the future. Explicitly signaling that policies will undergo change may impede enforcement, make decisionmakers seem indecisive, and make it easier for them to succumb to political pressure from special interests. Stakeholder groups, such as trade associations, consumer representatives, and environmental advocates, may prefer stability rather than repeatedly contesting policy choices, especially when such groups fear that they may lose influence in the future.

Under these many pressures to maintain the status quo, directing and managing an orderly change process is difficult, which is why adaptive decision strategies are particularly salient. Rosenhead, one of the first to explore the connections among robustness, resilience, and adaptive strategies, defines a decision as a "commitment of resources that transforms some aspect of the decision-making environment" (Rosenhead, 2001, p. 186). A plan, which foreshadows "a set of decisions which it is currently anticipated will be taken at some time or times in the future" (Legey and Kazay, 2001, p. 71), is adaptive because it often also includes "an identification of an intended future state which necessarily implies a set of future decisions" (Legey and Kazay, 2001, p. 71). Walker et al. (2001, p. 284) similarly defines adaptive policies as comprising "sequential combinations of policy options. Some options are to be implemented right away; others are designed to be implemented at an unspecified time in the future, or not at all if conditions are inappropriate."

Based on such definitions and adapted from McCray et al. (2010), an organization committed to planned learning will thus demonstrate

- a prior commitment to subject its existing plans to review and adjustment, along with fostering institutions, norms, and culture that facilitates such adjustment
- systematic effort to mobilize new factual information to inform the adjustment of the plan
- active consideration of the opportunities for such future adjustment and the potential for future learning in the choice of current actions.

Essential components of an adaptive plan include a planned sequence of actions, the potential to gain new information that might signal a need to change this planned sequence, and actions that would be taken in response to this new information. The sequential decisions of traditional decision analysis (Morgan et al., 1990), as well as real options approaches (Trigeorgis, 1996), follow this structure. The concept of "adaptive management" also follows this pattern. First introduced by C. S. Hollings (1978) to describe near-term policies specifically chosen as experiments to generate scientific understanding, adaptive management has since been used more generally to refer to policies that are designed to respond to new information— that is, those that incorporate planned learning.

Dewar (1993, 2002) describes a planning methodology in which decisionmakers identify the key assumptions underlying a proposed plan. Dewar then defines shaping actions as decisions that aim to make key assumptions more likely to hold, hedging actions as decisions to be taken if key assumptions begin to fail, and signposts as observed events or thresholds that suggest that such an assumption is indeed failing. RDM, which builds on assumption-based planning, aims to facilitate the development of adaptive management strategies and, in particular, combinations of shaping actions, signposts, and hedging actions that comprise these strategies (Lempert et al., 2006; Groves et al., 2014; Bloom, 2015). Box 2.1 provides an example of this process. All of these approaches depend critically on key indicators that signal the need for reevaluation of plans, redirection of effort, or new strategies.

Box 2.1. Signposts for Adaptive Management

The Metropolitan Water District imports water into semiarid Southern California and also helps manage the local supplies that provide about half of the region's water. The agency's 2010 Integrated Resource Plan Update (IRP) describes a preferred mix of alternative water supplies designed to meet the agency's goals through 2035, including reliability, cost, and environment. Recognizing that conditions are highly uncertain, the IRP also describes an adaptive management approach to monitor key trends and guide any future adjustments to the IRP.

To help implement this adaptive management approach, Metropolitan used an RDM vulnerability analysis to distinguish the specific future conditions in which the IRP is likely to meet or miss its goals. The results of this analysis, shown in Figure 2.4, indicate that the two most important uncertainties affecting the IRP's success are demand (horizontal axis) and supply (vertical axis). (Other trends considered in the analysis, but not shown on the axes, are less important for Metropolitan to monitor.) The supply indicator is a combination of the effects of climate change on precipitation and the effectiveness of Metropolitan's investments in groundwater management. In the red region—the high demand, low precipitation scenario—the IPR misses its goals.

Metropolitan developed its IRP assuming conditions in 2035, as shown by the large dot in the figure. The agency can monitor conditions as they evolve in the future, and, if they stray toward those of the high demand, low precipitation scenario, the agency may have to adjust its plans.

Figure 2.4
Example of Vulnerability Analysis

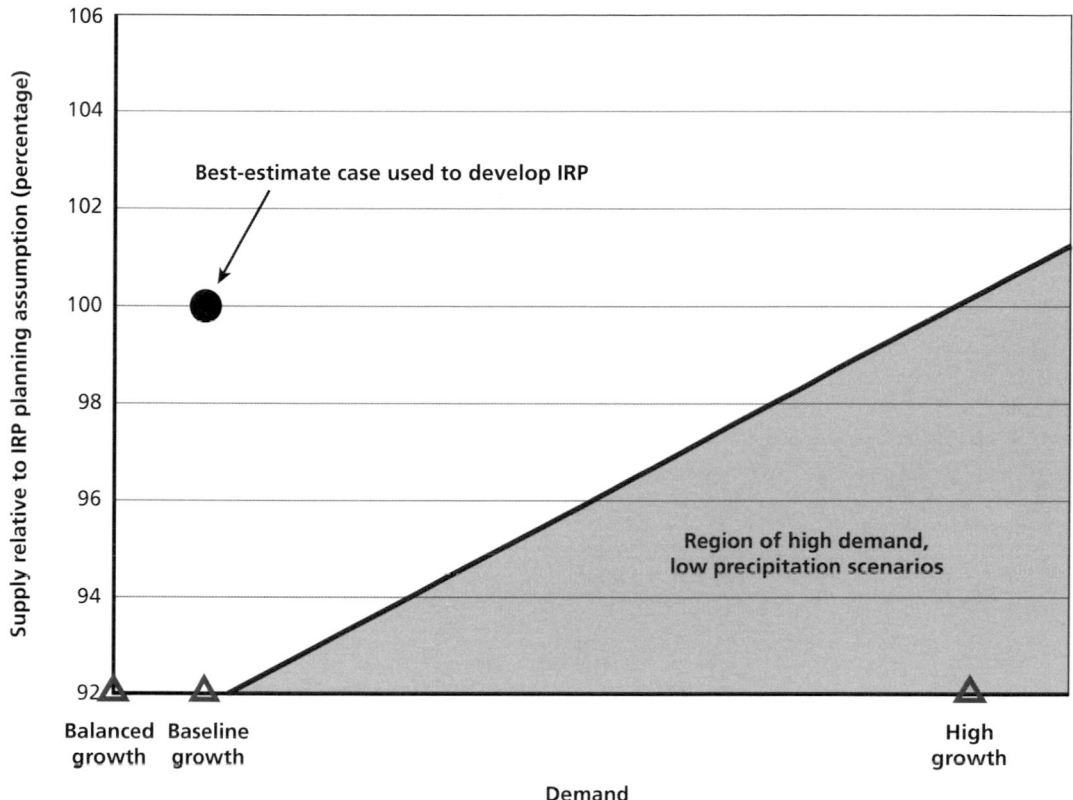

SOURCE: Parker et al., 2015. Used with permission.
RAND RR1144-2.4

Transformational Change

Climate change presents a particularly difficult challenge of transformative collective action. Both limiting the magnitude and adapting to the unavoidable impacts of climate change will ultimately involve some level of transformation, as defined in Table 1.2 (IPCC, 2013). Transformation generally requires significant changes by many different actors within a complex system. An urban area that is able to reduce its GHG emissions to zero, or significantly alter its land use patterns to manage flooding, would have had to undergo such a transformation.

Risk governance encompasses a wide range of actors, their institutions, political culture, and social norms. But the assessment of management options generally occurs within a decision-centric framework that isolates an actor within this interconnected system and examines the consequences of the choices available to that actor. This raises the question of who is a relevant actor. At one extreme, many climate risk studies take the viewpoint of a single global decisionmaker that can choose GHG emission reduction paths (Stern, 2007). Alternatively, climate risk management typically focuses on policy choices available to a single administrative department in a city government, or even individual residents. Both perspectives can be useful. In general, actors closer to the local scale are responsible for most tangible efforts toward climate risk management.

How can those involved with promoting urban climate risk management think about which actors could and must be involved in pursuing any needed transformational change? To begin to answer such questions, Figure 2.5 distinguishes among three tiers of processes that can affect the ability of urban areas to pursue climate risk management, from business-as-usual operations in the lowest tier to the most challenging transformation of governance arrangements through new laws, policies, or programs in the top tier. This figure draws on the "learning loop" framework (Kolb et al., 1975; Argyris and Schöen, 1978; Hargrove, 2002; Keen et al., 2005; Peschl, 2007; Pelling et al., 2008) often used in the resilience and climate change adaptation literature to conceptualize processes of learning and change.

Tier 1 represents actions that can be taken by existing departments in city government, existing groups, and collaborations among them, without changes in laws or any significant change in policy or budgetary allocations. Tier 2 represents actions that involve relatively minor changes in law or institutional structure at the local, state, or

Figure 2.5
Tiers of Transformation

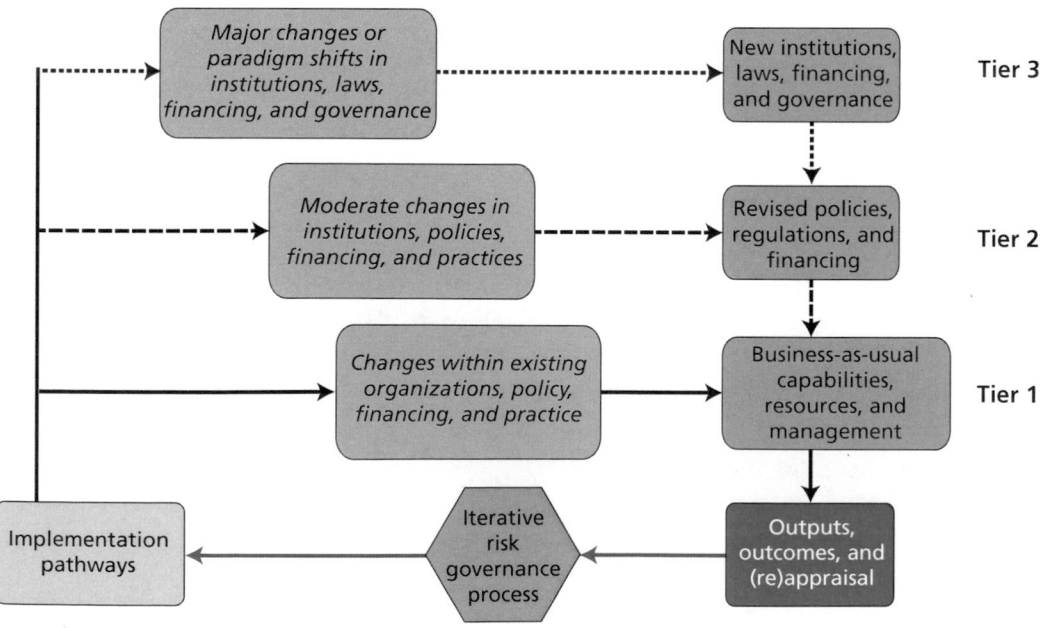

national level. Tier 3 involves more significant changes in laws, regulations, funding, and institutions at the state and national levels in which the cities operate; the economic environment; and the broader social and political context, including, for instance, trust in government, the authority of science, and degree of civic involvement.

For example, a city's efforts to reduce its own GHG emissions through investments in renewable energy and conservation would fall in Tier 1. State or national renewable energy standards might fall in Tier 2. State and national carbon taxes might fall in Tier 3. From an adaptation and resilience perspective, a city's efforts to adapt to a drier climate by incentivizing drought-tolerant landscaping would fall in Tier 1. Facilitating the use of wastewater recycling by revising state health codes would fall in Tier 2, as would the inclusion of water reliability in the rating of municipal bonds. As a real-world example of Tier 1 changes, several departments in the City of Los Angeles launched the One Water LA 2040 initiative to engage citizens in long-term planning for the city's water, wastewater, and storm water systems. Thus far, activities fall within existing organizational and management arrangements, but as the stakeholder engagement and planning processes advance, Tier 2 and possibly Tier 3 strategies may emerge as vital to achieving the city's goals (City of Los Angeles, 2014).

Climate risk management strategies that rest on actions in Tier 1 are preferable from an operational perspective to those in the other tiers because specific local actors are often already empowered to take action, and those actions are often easier to implement. But actions in Tier 1 tend to be more limited in scope and impact than those in the upper two tiers. A city and region will generally prove more resilient to the extent that it is organized to take a wider range of actions within Tier 1. Such capabilities will depend on the extent to which the jurisdictions and incentives within which they operate (created by Tier 2 and 3 changes) are well-aligned with the challenges and response options related to climate change. Currently, many jurisdictions have not managed to reach beyond their status quo Tier 1 authorities and processes. Nonetheless, modest changes at this level could potentially build a public case for more effective climate risk management strategies that could be achieved through transformations of the kind envisioned in Tiers 2 and 3.

Change processes in these tiers pose several challenges that are related to the concept of transformation. Changes in Tiers 2 and 3 generally involve coordinated efforts by many actors and, therefore, are harder to organize. Climate change alone may not be seen as sufficient reason to undertake the magnitude of change contemplated in these upper tiers, Tier 3 in particular, so climate considerations will generally need to be incorporated into other policy efforts, perhaps related to energy, transportation, or taxation. Thus, those pursuing climate risk management strategies may be particularly interested in synergies between climate and other goals, as well as in actions available in Tier 1 that can incentivize or catalyze appropriate changes in the upper tiers.

Decision support methods and tools can help decisionmakers understand, design, and implement transformative strategies. For instance, the vulnerability analysis

(Step 3) of an RDM decision support process, shown in Figure 2.4, can identify the range of conditions that can be managed and the risks that can be reduced by actions within each of the tiers of Figure 2.5. Similarly, such analyses can suggest how changes in Tiers 2 and 3 can increase the range of conditions managed within Tier 1. Using the language of adaptive management, one would like a larger rather than smaller range of risks to be managed within Tier 1. An appropriate vulnerability and response option analysis can suggest the actions in the upper tiers that can most effectively expand the coping range in Tier 1's business-as-usual operations. Incorporating concepts from game theory, such analyses can also suggest how actions in Tiers 1 and 2 might catalyze effective changes of the kind envisioned in Tier 3 (Isley et al., 2015). Network analysis could also reveal the current connections among actors and how interventions in Tier 1 might affect the networks required to affect Tier 2 and Tier 3 changes.

In the next chapter, we combine the governance and transformational change concept in Figure 2.5 with iterative risk management (Jones et al., 2014)—in particular, the RDM process from Figure 2.3. Outcomes inform the scoping process and help suggest whether and what transformation beyond business as usual in Tier 1 might help achieve various goals.

Combining the tiers of transformation with an RDM-based deliberation with analysis process provides the context for determining how and where current practice in each tier may create or expose vulnerabilities and what types of actions might reduce those vulnerabilities. For instance, an RDM-based iterative risk management analysis would produce vulnerability maps that address such questions as the following: Over what range of futures does the status quo or proposed alternative plans meet risk reduction, cost, and other goals within the current Tier 1 structure? Over what range of futures would changes in Tiers 2 and 3 expand the ability to meet such goals? Such results could then support exploration of the types of actions within Tier 1, as well as actions that require the higher demands of transformation represented by Tiers 2 and 3. This structure is generally consistent with the three-layer city resilience framework of da Silva and Morera (2014) that links urban infrastructure with the dynamics of urban governance, as discussed in Appendix C.

As we noted at the beginning of this chapter, we seek indicators to associate with the risk governance framework that are tailored components of specific decision support and decisionmaking processes, as well as the urban climate response strategies that arise from these decisionmaking processes. These indicators describe the system as it exists; the capacities within the system to enable change; and the near-, intermediate-, and longer-term outcomes of the change and transformation processes. In the next chapter, we present a structure for these indicators to support the risk governance framework and then consider which existing systems of indicators may be relevant. Most of these indicator systems were developed for narrower program and project evaluation purposes, but some include indicators suitable for supporting decisionmaking for both mitigation and adaptive responses to climate change in urban areas.

CHAPTER THREE

Indicators to Support Climate Risk Governance and Decisionmaking

In Chapter Two, we proposed that a risk governance decision-making framework, augmented with insights from other frameworks relevant to decisionmaking in complex systems, is well-suited for urban climate risk management. In this chapter, we seek to construct a system of indicators consistent with the multiple dimensions of risk governance. We begin with a brief overview of the relevant literature on indicators, with details provided in Appendix C. We then lay out an approach consistent with the risk governance framework and present a hypothetical example of how the framework and indicators might work in practice. Our intent is to identify indicators that can support place-based decisionmaking and thus enable urban areas to improve their own climate risk management. By necessity, each urban area will need to draw on data sources and topical indicators most appropriate for its own particular circumstances. Nonetheless, we expect there to be commonalities among both the indicators and the data sources populating the indicators across different cities.

Challenge of Developing Decision Support Indicators

Climate-related indicators have been developed for many purposes and contexts. As summarized in Appendix C, indicators have been developed by or for international development agencies to evaluate the effectiveness of their projects and project portfolios and to assess the quality of planning by recipients of development assistance to inform future resource allocation decisions by the development agencies. Other indicators have been developed specifically to evaluate progress and the quality of planning for adaptation or resiliency by cities, regions, and countries. Some of these indicators are focused on process; others encompass both process and outcomes. Indicators may be quantitative or qualitative, the latter often capturing public perceptions that can influence the political calculus of a complex decision with winners and losers.

While we are most interested in indicators designed specifically to support deliberative decisionmaking processes, we are nonetheless interested in building on a range of existing approaches that may still be useful for our purposes. However, the variation in purpose, place, and sector does make a difference, making it difficult to develop meaningful indicators that apply equally well in different contexts and across regions and sectors (Bours et al., 2013). For this reason, many proposed indicator systems found in the scholarly literature offer general frameworks intended to be customized for particular applications.

The emerging practice and scholarly social science literature of indicators to support climate-informed decisionmaking, whether for measuring reductions in GHG emissions or measuring effectiveness of adaptation and resilience strategies, draws on well-established systems of indicators developed in the business and international development assistance literatures. In the application of these methods to adaptation studies, leading researchers (Preston et al., 2009; Brooks et al., 2011; Preston et al., 2011; Moser et al., 2013) have been moving in three important new directions: (1) from a focus on such programmatic inputs as funding or high-level plans to a focus on concrete measures of success; (2) from a focus on adaptation as a stand-alone endeavor to a focus on addressing climate risk as one of many activities conducted by governments, the private sector, and other organizations; and (3) from a focus on estimation of the vulnerability of specific populations and sectors to a focus on reduction and management of risks.

The adaptation indicators literature is grounded in the broader literature on program and project evaluation. In particular, the use of models for program evaluation has become widespread in the philanthropic sector, and proposed indicators for climate adaptation draw on these ideas. Logic models provide a causal representation of the ways in which program inputs (e.g., resources, activities) are related to the desired outcomes. Indicators can then be linked to the nodes on this causal chain. For instance, if a social welfare organization aims to provide meals to the needy, a logic model might suggest that the organization needs to gather sufficient food, assemble volunteers to distribute it, obtain a space to serve it, and so forth. Indicators could be linked to each of these steps, as well as the nutritional gains of the community served. As described later in this chapter, several proposed climate adaptation indicators draw on this logic model framework to help define various categories of inputs and link them to desired outcomes.

The climate adaptation indicators literature draws less explicitly on other frameworks, such as balanced scorecards (Kaplan and Norton, 1992; Kaplan, 2010) and the ISO total quality environmental management standards (e.g., ISO 14001), often used by businesses and large government organizations (ISO, 2015; National Research Council, 1999). For example, the balanced scorecard literature helps organizations move beyond a narrow focus on financial performance to a broader consideration of multiple objectives, such as their environmental and social impacts. Scorecards aim to align an organization's full set of strategic goals with their internal processes, investments to develop future capabilities, and demands for current and future resources. In their process role, balanced scorecards are well-suited to helping distinct but related organizations to act cost-effectively in concert, with the added advantage of having large communities of practice. This literature also provides a means of rolling up many hundreds of potential indicators to a small number and of aligning the indicators used by different parts of complex organizations. Despite the lack of a specific link, some proposed climate adaptation indicators do draw on these ideas by suggesting a small number of indicators to help organizations align their internal processes and investments with their strategic goals.

Proposed evaluation frameworks for climate adaptation and mitigation face a particularly difficult challenge because there is often significant temporal distance between inputs and outputs and an indirect and uncertain connection between outputs and outcomes. Most obviously, an urban area may launch efforts to eliminate its emissions of GHG (an output of emissions reductions strategies), recognizing both that it may take many decades to reach such a goal and that the ultimate consequences of doing so—helping to limit the magnitude of future climate change (a suite of outcomes)—will depend on whether or not the rest of the world also meets a stringent emission reduction goal. Similarly, an urban area may take steps to reduce its exposure and vulnerability to climate-related disasters (outputs) in anticipation of future changes in the frequency of extreme events, but it (fortunately) may have limited opportunities in any given decade or two to observe the quality of these preparations. The outcomes of these steps would ultimately be the enhancement of economic and social well-being and the preservation of lives and property.

Thus, consistent with the broader literature on program and project evaluation, many proposed evaluation frameworks use proxy indicators of adaptation preparedness, processes, plans, projects, and programs—all aimed at eventually realizing successful adaption or mitigation outcomes but for which outcomes cannot be directly measured, at least in the near term. Most of the indicators and frameworks that we reviewed fall under this category, and for good reason. The focus on measuring outputs of processes rather than measuring how adaptation and mitigation efforts avoid future climate change impacts (outcomes) is a reflection of how hard the latter is to do:

> While often treated as a gold standard in the general monitoring and evaluation literature, [outcome-based indicators] have not been widely used in an adaptation context, reflecting the difficulty of attributing reduced impact specifically to adaptation, where success may not be apparent for decades and where impacts averted in the future are tricky to estimate (Ford et al., 2013, p. 3).

Most examples of adaptation indicators come from the gray literature (outside of peer-reviewed scholarly journals) and have been developed by governments, international development organizations, and NGOs as they seek to track their projects and investments. Without widely accepted indicators or a strong academic literature from which to draw, Leclerc notes in the context of international development activities that "the organizations responsible for implementing and financing adaptation have had no choice but to try and test indicators for reporting back on how international funds are being used" (Leclerc, 2012, p. 5).

Table 3.1 presents a useful typology of current climate adaptation tracking approaches developed by Ford et al. (2013, pp. 4–5) and distinguishes outcome-based

Table 3.1
Outcome-Based Versus Process-Based Approaches to Adaptation Tracking

Tracking Approaches	Characteristics	Data Sources	Strengths	Limitations
Outcome-based	Outcome evaluation: *reduced negative climate change impacts*	• Track climate-related losses, mortality, and morbidity, over time and in relation to adaptation • Examine impacts of climatic hazard event before and after adaptation	• Natural hazard loss databases (e.g., emergency events database)	• Quantification of adaptation progress and effectiveness • Metrics can be monitored over time • Availability of standardized global datasets of hazards losses and mortality across regions • Legitimacy within policy evaluation community	• Applicable only where outcomes are directly observable • Difficulty of inferring causality between outcome and adaptation • Potential for maladaption not captured • Limited applicability to "soft" and mainstreamed adaptations • Does not measure outcomes from adapting to wider (nonevent-oriented) climate change
Preparedness-, process-, and policy-based	Adaptation readiness: *presence of key governance factors essential for effective and successful adaptation*	• With regard to adaptation, evidence of: political leadership; institutional organization; stakeholder involvement; climate change information; appropriate use of decisionmaking techniques; and consideration of barriers to adaptation, funding, technology development, and adaptation research	• Speeches at Conference of the Parties meetings • Attendance at Conference of the Parties meetings • Leadership identified in UNFCCC National Communications or National Adaptation Programmes of Action • UNFCCC National Communications • National assessments	• Not dependent on outcomes being visible • Captures readiness for future action and ability to effectively implement adaptations	• Need to validate if readiness translates to action • Limited availability of readiness metrics

Table 3.1—continued

Tracking Approaches	Characteristics	Data Sources	Strengths	Limitations
Process-based approaches: *process through which adaptations are developed and implemented in pursuance of a desired outcome or objective*	• Comparison of adaptation characteristics and steps of development to theoretically and empirically derived characteristics of adaptation success and best practice	• National Adaptation Programmes of Action • Adaptation inventories	• Not dependent on outcomes being visible • Capture the key processes that are believed to underpin effective and successful adaptation	• Limited systematically collected data on process of adaptation development and implementation • Limited transferability across nations • Time intensive • Unproven link to adaptation success
Analyzing policies and programmatic approaches: *monitoring and comparison of reported adaptation actions and their characteristics*	• Analysis of characteristics of reported adaptations and comparison across regions, by vulnerability categories, over time, and with respect to adaptation "obligations"	• UNFCCC National Communications • National Adaptation Programmes of Action • Adaptation inventories • National adaptation assessments	• Not dependent on outcomes being visible • Systematic and quantitative analysis of progress • Comparability across nations • Suited for global application • Amenable for rapid assessment	• Success not directly measured • Results subject to reporting bias
Examining measures of changing vulnerability: *measurement of change in vulnerability in relation to adaptation*	• Monitor aggregate vulnerability indexes in relation to adaptation actions • Focus on specific indicators which capture the generic determinants of vulnerability (e.g., access to education, poverty, health, and inequality) • Examine specific components of sensitivity and adaptive capacity to climate change impacts	• Climate Change Vulnerability Index • Environmental Sustainability Index • Global Climate Risk Index • GAIN Index	• Not dependent on outcomes being visible • Readily available vulnerability indexes globally • Amenable for rapid assessment	• Inability to capture determinants of vulnerability • Fundamental disagreement between indexes on magnitude of vulnerability • Challenge of linking change in indexes to adaptation

SOURCE: Ford et al., 2013, pp. 4–5. Used with permission.

indicators (related to reducing negative effects of climate change) from process-based indicators (related to outputs that may lay a groundwork for eventual achievement of outcomes). Our proposed indicators, described later in this chapter, reflect the Ford et. al. typology of outcome- and process-based (output) indicators but broaden Ford's definition of outcomes to include indicators of changing vulnerability also. Such vulnerability indicators could be considered intermediate outcomes rather than proxy indicators for final outcomes.

The literature on indicators associated with mitigation activities also distinguishes between policy inputs and policy effects. Our proposed indicators draw on a recent review by Singh and Vieweg (2015, p. 5) summarized in Table 3.2. Singh and Vieweg offer a comprehensive overview of potential outcome-oriented indicators associated with intermediate policy effects (output) that relate to buildings retrofitted with more energy-efficient equipment; GHG effects that relate directly to measured reductions in emissions (output); and non-GHG effects that relate to reductions in air pollution (output), improvements in health (outcome), or income growth (outcome). Singh and Vieweg (2015, p. 8) make brief mention of "transformational" indicators associated with policy change but generally restrict their definition of transformation to structural changes within economic sectors and technological change. They do not mention transformation in governance explicitly, nor do they explicitly discuss the use of indicators in a decision support process, although they note the value of these indicators for policy design (as distinguished from policy choice). Their primary focus is on performance measurement and tracking of policy implementation.

Table 3.2
Indicators Associated with Mitigation Policies

Policy Implementation Indicators		Policy Effects Indicators		
Input Indicators	Activity Indicators	Intermediate Effects Indicators	GHG Effects Indicators	Non-GHG Effects Indicators
• Finance • Human and organizational resources • Other inputs	• Licensing, permitting, and procurement • Compliance and enforcement • Other policy administration activities	• Behavioral changes • Technology changes • Process changes	• Changes in GHG emissions	• Changes in environmental, economic, or social conditions, other than GHG emissions changes
• Changes in indicators directly related to policy implementation • Data often available from the entity implementing the policy		• Changes in indicators may additionally be influenced by factors beyond the policy • Changes are likely to be observed in the target group(s) of policy, making data collection more challenging		

SOURCE: Singh and Vieweg, 2015; CC BY 4.0.

Relevance of Existing Indicator Systems

Measuring adaptation and mitigation progress concerns a variety of institutions and organizations. Governments, NGOs, and academics have developed many frameworks aimed at measuring climate-relevant policies at different geopolitical scales and with foci on adaptation plans, processes, and projects. Based on our review, however, the indicator frameworks have a number of common limitations. First, most frameworks—with the possible exception of frameworks developed by and for the international lending and aid organizations for evaluating adaptation projects and portfolios—have not been widely applied, for a variety of reasons. Some appear to be one-off academic exercises, while others are still under development and pilot testing. As such, it is difficult to assess the utility and applicability of these indicators in practice.

Second, many indicator frameworks seem to lack strong theoretical underpinnings and methodological rigor. In many cases, the frameworks are *ad hoc* and emerge from governments' and NGOs' practical and urgent needs for assessing investments and measuring progress. Two adaptation metric frameworks, Tracking Adaptation and Measuring Development (TAMD; IIED, 2013) and Preston et al.'s (2011) framework, seek to overcome this shortcoming with explicit attention to methodology.

Third, as with measurement instruments in other contexts, it is important to know whether a set of indicators is reliable and valid. In the climate change adaptation context, a reliable set of indicators would produce the same assessment of progress on climate change adaptation under the same or similar set of conditions. Validity is especially important: Does a city's success as measured by the indicators actually translate into climate resilience and the aims of climate change adaptation efforts actually being met? Singh and Vieweg (2015) pay particular attention to measurability, replicability, and data quality issues in the context of mitigation indicators.

Finally, many frameworks in our review treat adaptation and mitigation as largely linear processes, from recognizing climate change as a concern to implementing adaptation and mitigation actions. Simple frameworks may be easier to understand, apply, and evaluate. They may resonate well with stakeholders. However, they may also miss important and interwoven complexities of urban transformation, organizational behavior change, the interplay between politics and science, and governance changes, which may shape actual climate change responses. Nonetheless, it should be noted that starting with a simple framework is a reasonable path toward developing buy-in among stakeholders early in a deliberative process and eventually could lead to more sophisticated, nuanced approaches for both systems models and indicator frameworks.

In general, the indicator frameworks we reviewed do not make this trade-off between simplicity and richness explicit. Most do not discuss such complexities at all. Preston articulates these concerns with regard to adaptation:

> Adaptation plans frequently mention the need to mainstream adaptation into existing policies and capitalize upon synergies among adaptation and other policy

goals. Yet, adaptation plans themselves largely frame adaptation in a narrow, climate-centric manner that overlooks the capacity and institutional challenges associated with the process. In addition, the simple fact that institutions have opted to develop management plans that are specific to climate change, and in some cases solely adaptation, rather than integrating climate change into the range of existing policies and environmental management efforts, suggests that demonstrating action on climate change (as opposed to securing positive societal and ecological outcomes) may be the key driver of adaptation planning in many instances (Preston et al., 2011, p. 428).

To date, we cannot consider any particular framework "mature" or "validated"—most are under development, and only a few have been applied in the field. In sum, the existing frameworks for indicators for climate change responses are not particularly well-suited to support decisionmaking in complex governance settings.

Framework for Indicators Within a Decision Support Process

To show how indicators could be integrated into a decision support and generalized risk governance process, we begin by proposing five general types of indicators for urban climate risk management. As shown in Table 3.3, these indicators are divided into two overall categories, those relating to the state of the urban region and those related to the quality of the plan and the process that produced it. Figure 3.1 shows

Table 3.3
Proposed Categories of Indicators

Indicator Categories	Key Attributes
State of the Urban Region (Based on Causal Models)	
Well-being and risk to well-being	• Objective and subjective measures of real and perceived conditions among residents and the community at large
Capacity to implement the plan	• Management capacity • Technical skills
Capacity to adapt and transform	• Political will • Public acceptance and support • Adaptable governance mechanisms • Sustainable financial resources
Quality of the Planning	
Quality of the planning process	• Inclusivity • Legitimacy • Analytical grounding • Cost-effectiveness
Quality of the resulting plan	• Well-structured • Designed for evaluation and learning

indicator categories related to the former in green boxes and those related to the latter in blue. Figure 3.1 is identical to Figure 2.5 but with the addition of the iterative risk management process and the categories of indicators shown on the graphic. The effect of current actions on changes in the tiers—the consequences of those changes—can be addressed with various types of models, as discussed later in this section.

Well-Being and Risks to Well-Being Indicators

Well-being and risks to well-being refer to conditions and risks experienced by people in the city now and in the future. Current conditions—such as income, health, access to recreation, and biological diversity—and some current risks—such as tidal flooding—

Figure 3.1
Enhanced Iterative Risk Management Process with Indicators

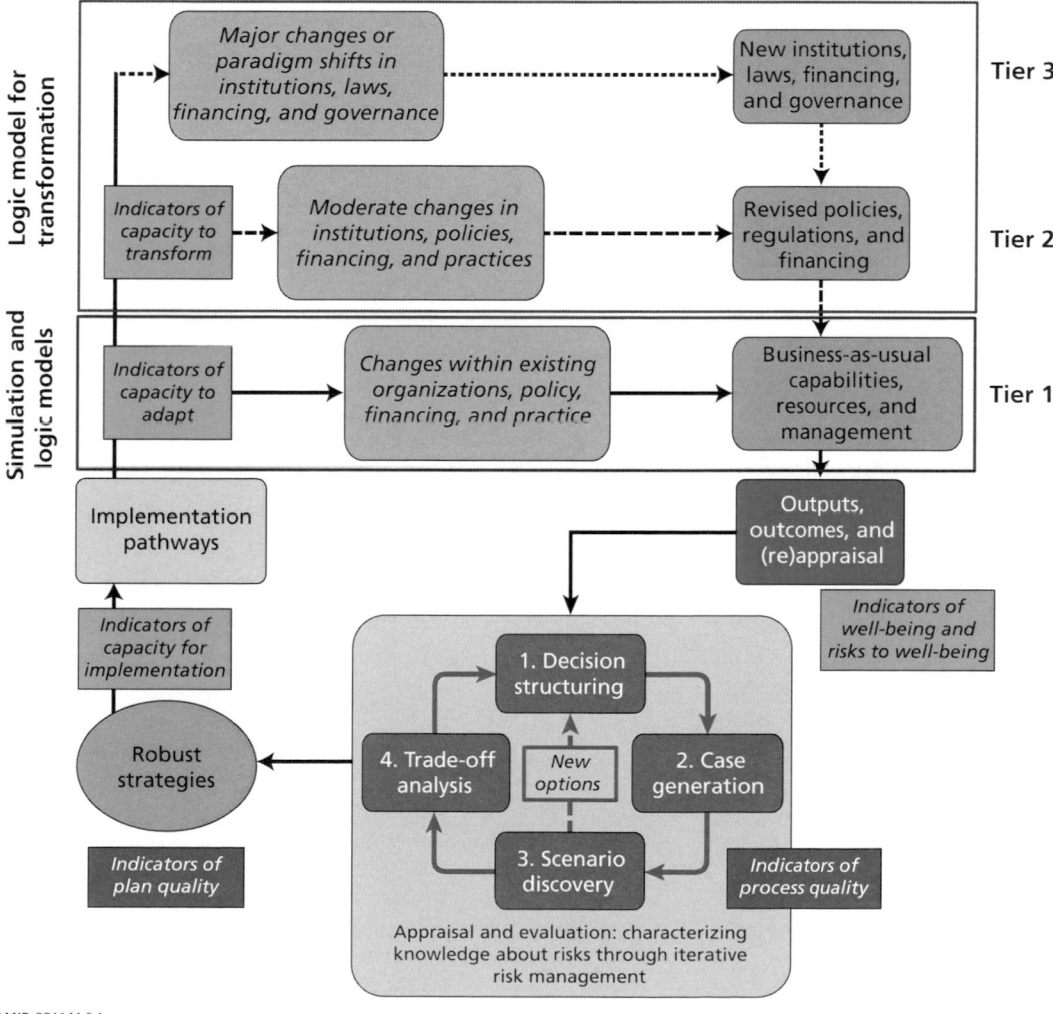

can often be measured directly. Indicators can be drawn from such measurements. In general, however, urban climate policies have important impacts in the near to far future. Such future conditions and risks can only be estimated using some type of mathematical model. These models are inherently error-prone because of the complexities of the underlying systems and the difficulty of translating that complexity into mathematical statements. However, they have the virtue of imposing a structured and transparent approach to the estimation process, using the best available science. Indicators can then be drawn from the outputs of such models.

Estimating future conditions and risks is, however, complicated for at least two reasons. First, calculating future risk requires estimating future hazard, exposure, and vulnerability. In the early days, climate impacts analysis often only considered future hazard by superimposing projections of future climate on current socioeconomic conditions. This is clearly wrong, since changes in exposure and vulnerability can significantly increase or reduce climate-related risk (Jones et al., 2014). Thus, more credible estimates of future risk require projecting not only future climate, but also how an urban area may evolve.

Second, to be useful as part of a decision support process that aims to inform the design of urban climate risk management plans, future risk estimates require estimating how today's actions will influence that risk. In some cases, this is relatively straightforward. For instance, contingent on projections of exposure and vulnerability, one can use simulation models of physical systems, based on well-established scientific principles, to estimate future risk reduction from storm events (a hazard) that would result from building a levee of a certain size. In other cases, the process is less straightforward, because today's climate policies also aim to influence future exposure and vulnerability. For instance, a climate risk management plan may seek to catalyze Tier 3 changes that significantly alter incentives and norms for a city's residents and businesses, influencing where people choose to live and the type of structures in which they live, thereby significantly reducing future exposure and vulnerability. Knowledge of that chain of causality in theory and measuring it in practice, however, is another matter, and it often proves difficult to estimate the extent to which such policies achieve their goals.

In many cases, and in most of the cases with which the authors and their colleagues have been involved, there exist quantitative and validated simulation models for many biophysical and socioeconomic systems relevant to urban climate risk management.[1] As shown in Figure 3.1, these simulation models generally represent systems

---

[1]   While there continues to be controversy in the physical sciences about the criteria for assessing the validity of a simulation model, the general principle is that the model can be shown to reproduce a wide range of historical conditions using a different data set from the one initially used to calibrate the model parameters. The most difficult aspect of validity to establish is the underlying structure of the model itself, and some level of model error must be assumed to exist in nearly all simulation models of physical systems. This is particularly true if such models are used to extrapolate past the range of observed data, as is often the case in multiscenario analyses. Exploratory modeling (Bankes, 1993), which underlies the RDM approach suggested here, provides a framework for managing such parametric and model uncertainty in decision support applications.

at the Tier 1 level. The outputs of such models can be used as proxy indicators for future well-being and risk. For example, hydrologic, rainfall runoff, and water quality simulation models are available in many jurisdictions, and these can be modified to relate investments in gray and green infrastructure to water quality and flooding in each district of a city under various climate scenarios. In such models, the investments are policy inputs. The resulting projections of pollution loadings and flood depths, when combined with projections of the geographic distribution of the region's population, represent model outputs that can be used as indicators of future health and flood risk. As another example, travel demand models, used widely by metropolitan planning organizations, relate investments in transportation infrastructure (inputs) and projections of future land use, population distribution, and economic activity to such outputs as traffic flows and miles traveled. These model outputs can then be converted to proxy indicators of future mobility and GHG emissions.

Capacity to Implement Indicators

The *capacity to implement* the plan refers to the current resources and context available to the planners. Temporally, indicators of the capacity to implement refer to today's ability to carry out the plan and to generate outcomes that might be observed in the near term. This is a critical determinant of successful policy and includes such issues as availability of financing and funding; technical and administrative capacity within the region to manage complex design, construction, and maintenance of plan components; and sustainment of public support for the plan's implementation.

Capacity to Adapt and Transform Indicators

The *capacity to adapt and transform* refers to the urban region's ability to influence its future conditions and capabilities. The IPCC (2014c) defines *adaptive capacity* as the "ability of systems, institutions, humans, and other organisms to adjust to potential damage, to take advantage of opportunities, or to respond to consequences."

What we call *capacity to adapt indicators* reflect such capabilities in an urban region, but we use the term more broadly to refer to both a capacity to address the impacts of climate change as they unfold and a capacity to successfully adjust policies that reduce GHG emissions as challenges and opportunities arise in the future. An urban region's capacity to adapt is limited by many fundamental attributes of the system, including its institutions, laws, human capital, and culture. The capacity to transform represents the ability to relax these constraints and open up opportunities by significantly changing such attributes. The capacity to implement represents the ability to carry out the near-term steps of an adaptive management strategy. The capacity to adapt represents the ability to monitor and adjust course as needed in the future. The capacity to transform represents the ability to alter the institutional, legal, and other constraints on the future capacity to implement and to adapt.

Estimating the capacity to adapt and transform involves judgments about how current conditions and actions may affect future capabilities and, in general, requires some type of model. But quantitative, phenomenological models may not often exist for such purposes. In such cases, employing more qualitative logic models may be the most effective recourse. Logic models can derive from several sources. Most simply, they can be elicited from local decisionmakers, who may have insight about how their actions and the actions of others contribute to the capacity to adapt and transform. In other cases, sufficient social science data and theory may exist to fashion one or more science-based models of how the actors in the system could contribute to these capacities. Either or both of these sources could contribute to an appropriate logic model (and in some cases, the latter could contribute to a causal mathematical model) that could be used to relate policy inputs to estimates of outputs and ultimate outcomes. As shown in Figure 3.1, simulation and logic models might both usefully represent systems at the Tier 1 level, but in general, logic models may be the preferred option for Tiers 2 and 3.

Many factors affect the capacities to adapt and transform, some of which may also be important in assessing the capacity to implement. For example, Renn (2008) identifies actor networks, political and regulatory culture, and social climate as foundational to the capacity of places to implement transformational plans. He does not, however, specify how these general attributes translate into measurable objective or subjective indicators. Hallegatte's factors (World Bank, 2013), such as alignment of incentives, access to appropriate information, cultural biases and behavior, and financial resources, are relevant to both implementation and transformation capacities.

In addition, Pahl-Wostl's (2009) work on transformative capacity as a type of learning, from which our tiers of transformation in Figure 2.5 is drawn in part, focuses on the salient aspects of governance that may affect a region's capacity to rise to the challenges of climate change response. She identifies a number of categories of interest: regulatory and legal institutions, as well as informal institutions that reflect social norms on tolerance for uncertainty; actor networks and their connections; multilevel interactions among jurisdictions and leaders; and modes of governance. However, she does not define indicators of the functionality of any of these features.

Figure 3.2 offers an example of a logic model that might be used to suggest indicators for the capacity to transform, based on some of the factors identified by Pahl-Wostl. This example is drawn from a recent long-range plan from a metropolitan planning organization (MPO) responsible for transportation and land use planning in California (Association of Bay Area Governments and the Metropolitan Transportation Commission [MTC], 2013). This MPO argues that various changes in state law and in the state's constitution would improve its ability to reduce GHG emissions regionally and produce a better-connected and more efficient network of transportation options. Such changes clearly fall in the Tier 3 category, among the most difficult

Figure 3.2
Example of Logic Model for a Tier 3 Transformation Process

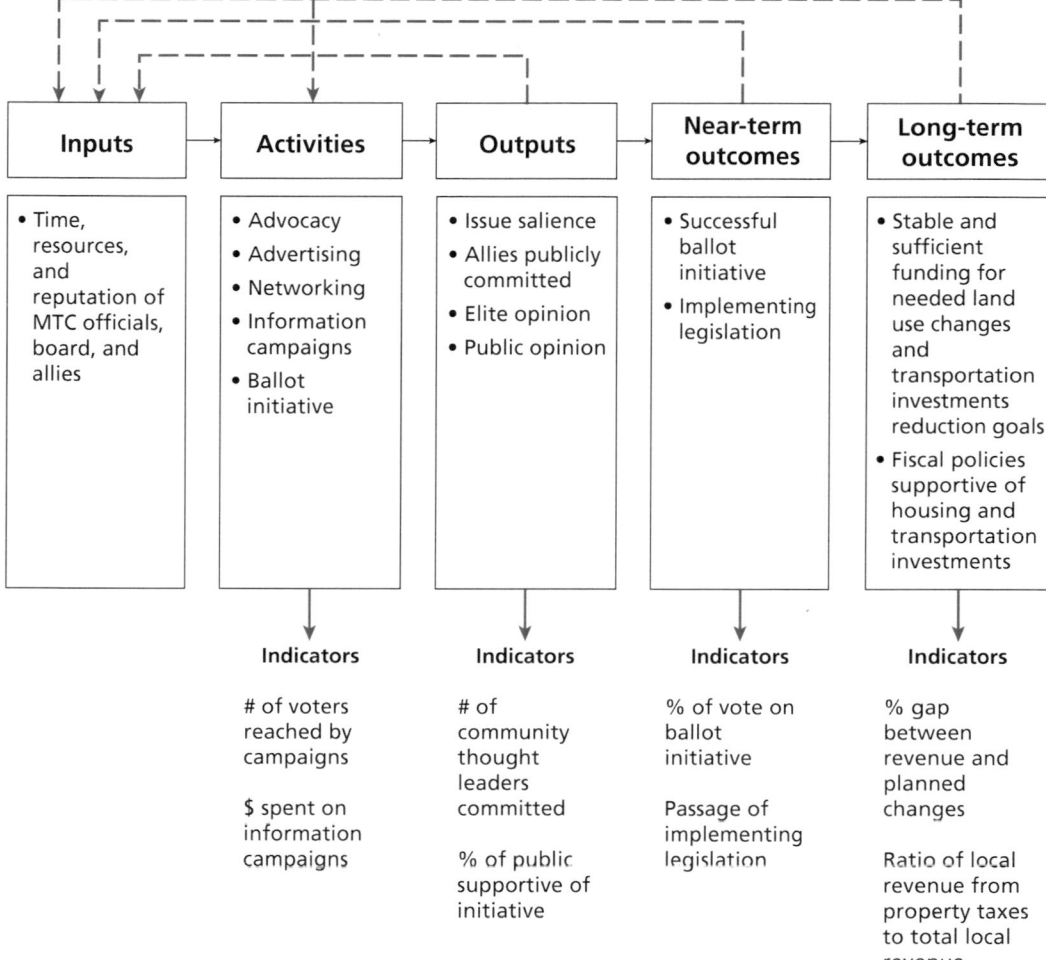

SOURCE: Created by the authors based on concepts from Association of Bay Area Governments and MTC, 2013.
RAND *RR1144-3.2*

changes in public policy at the state level. The logic model in Figure 3.2 shows the kinds of actions mentioned in the MPO's plan, including

- changing the state constitution to require that new local transportation funding ballot initiatives receive only 55 percent of the vote rather than the current two-thirds now required

- lifting the state constitutional limits on local property taxes and allowing local governments to draw on a more balanced source of tax revenue for housing and transportation improvements
- modifying the state's environmental quality law to offer incentives for in-fill developments.

The logic model provides a platform on which to elicit expert judgments on the MPOs' capacity to transform—that is, to carry out the listed activities and have them result in the desired outputs and outcomes.

In developing indicators for the capacity to adapt and transform, it is useful to ask how much such capacity is needed. The tiers of transformation shown in Figure 3.1 suggest how the relationship between these two capacities and the state of the city indicators could be quantified. Using various models, we could identify the range of futures over which the region and jurisdictions within it can meet their goals using adaptive strategies confined to actions possible within Tier 1. Vulnerability maps, such as the one shown in Figure 2.4 and described in Box 2.1, might be used to display the results of such analysis. We could then identify the range of futures over which the region and its jurisdictions could meet their goals with various Tier 2 and Tier 3 changes. Such analyses would suggest what levels and types of risk reduction can be achieved by specific changes in each of the three tiers. This information would help to define the benefits of such changes. Note that our framework does not explicitly mention barriers, but this concept is reflected in assumed constraints within each of the tiers. Changes in the tiers are synonymous with reducing barriers, and the capacity for transformation represents an evaluation of the extent to which current actions can help bring about the desired changes.

Measuring institutional capacity to adapt and transform is particularly challenging. The Helm and Sprinz (2000) scale of regime effectiveness in Figure 3.3 is suggestive of one potential approach to such a measurement. While the Helm and Sprinz application relates to the effectiveness of international environmental regimes, their framework may also be appropriate for considering the effectiveness of coordination among organizations within a city and within an urban region by understanding (1) how incentives, information, barriers, and resources might affect the choices made by diverse sets of actors in urban settings; (2) how these choices increase or reduce risk; and (3) how alternative interventions might change actors' choices and the resulting risks. The approach depends on a identifying a single performance metric that relates to the effectiveness of the governance regime to accomplish its key goal—for example, the percentage reduction in flood risk in a region relative to present conditions.

The Helm and Sprinz effectiveness score compares the observed environmental performance of the relevant region with two counterfactuals. The left of the figure shows the *no-regime counterfactual*—that is, the environmental performance expected with no institutional coordination, as represented by the best the actors could do if

Figure 3.3
Measure of Regime Effectiveness

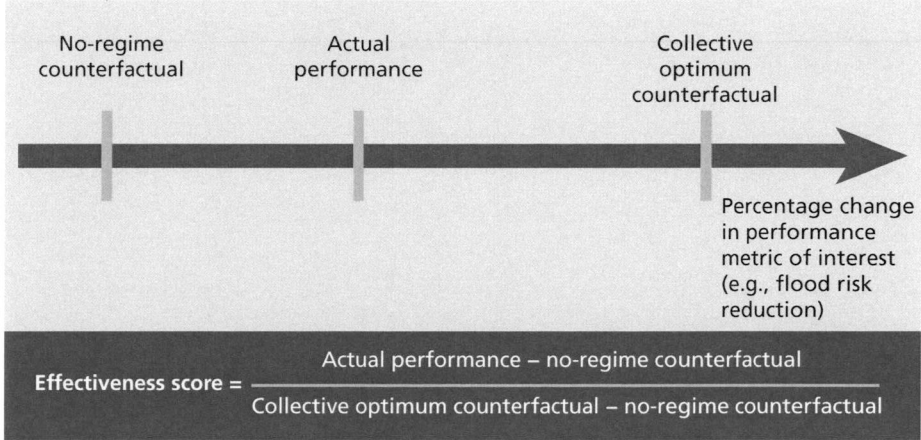

SOURCE: Adapted from Helm and Sprinz, 2000.
RAND *RR1144-3.3*

they pursued their self-interest without any formal coordination among themselves. The right side of Figure 3.3 shows the *collective optimum counterfactual*—that is, the performance that would result if all actors adjusted their performance to produce the social optimum. Helm and Sprinz measure the performance of the existing governance regime in relative terms, in which its actual performance lies between these two extremes. Risk management analysis simulations could estimate the level of risk operating within Tier 1 and the socially optimal level of risk reduction operating without any institutional constraints. The ratio of these performance levels would reflect the Tier 1 capacity, which in this example would be relatively low.

Quality of Plan and Planning Process Indicators

No set of simulation and logic models will ever provide a complete and infallible representation of the relevant urban system. (A well-known quip among modelers is that all models are wrong, but some are useful.) In addition, the literature on decision support makes clear that the process by which information is provided can be at least as important as the information itself in improving (or not improving) decisionmaking. For instance, the extent to which information is regarded as salient, credible, and legitimate may depend on how that information is generated and provided to stakeholders and decisionmakers. Thus, it is important to complement the objective and subjective well-being indicators with indicators related to the quality of the planning process and its products.

Plan quality refers to the attributes of the plan itself. Indicators of the quality of the plan and the processes that produce it are important because they are more closely tied to factors under the planners' control than are the outcomes produced by the plan.

In addition, many of the most important outcomes may be years in the future and can only be observed long after any evaluation of the plan would have occurred.

The decision sciences literature offers many taxonomies for good plans, but in general they focus on the extent to which a plan clearly articulates its goals, considers a wide range of alternative options, uses the best available science to estimate the potential consequences of those actions, considers a wide range of future contingencies, clearly confronts the trade-offs that exist among the plan's goals, and describes a credible approach for implementation and monitoring progress over time. A high-quality plan should also clearly assign responsibilities for actions and identify the resources required to implement the plan. For climate-related decisions, assessing the potential consequences of alternative actions requires a plan to be forward-looking—that is, to consider consequences stretching out into the future.

Preston et al. (2011) offer 12 types of indicators of plan quality in the categories of goal setting, decisionmaking, and implementation and evaluation. The UK Department for International Development's (DfID's) TAMD Track 1 indicators in Table A.6 also focus on how well climate and other relevant information have been factored into plans. The ability to evolve over time in response to new information will likely be important to any successful urban climate risk management effort. Table 3.4 provides a taxonomy that can help evaluate the extent to which such factors are included in urban climate plans, based on the work of Swanson et al. (2007; 2010). The first three elements refer primarily to attributes of the policies themselves. The third element, in noting that an adaptive strategy should often consist of a portfolio of policy actions that aim to influence a number of elements of a complex interacting system, draws on key insights from integrated water resource management and the concept of *defense in depth* employed in integrated flood risk management. The final four elements refer primarily to the context in which the policies are developed and implemented, which in general should encourage review and response, encompass a diversity of approaches to promote learning, decentralize decisionmaking, and emerge from a process of multi-stakeholder deliberation.[2] Note that the earlier elements in the taxonomy, such as *forward-looking* and *automatic policy adjustment*, tend to relate to information products that might be delivered by decision support tools for urban decisionmakers. The latter elements in the taxonomy, such as *multi-stakeholder deliberation* and *decentralized decisionmaking*, tend to relate to decision support processes and the institutions in which they are embedded.

Process quality includes events leading up to the present time in a planning process. These process indicators could consist of checklists of important attributes of

---

[2]   The taxonomy in Table 3.4 largely maps onto Swanson et al.'s (2007) seven tools for adaptive policies. But the latter taxonomy focuses on distinguishing between tools useful for anticipated and unanticipated changes. The former focuses more on distinguishing between attributes of the policy and those of the context in which they are developed. In addition, the latter taxonomy includes the attribute of integrated policies, which has become increasingly important in such areas as integrated water resource management.

Table 3.4
Attributes of Adaptive Plans

Attribute	Purpose
Attributes of plans	
1. Forward-looking	Identify longer-term vulnerabilities (including foregone opportunities) of near-term policy choices and potential responses to those vulnerabilities
2. Automatic policy adjustment	Specify signposts that indicate need for policy adjustment and contingent actions to take in response to those signposts
3. Integrated planning	Combine management of multiple elements of a system in a holistic plan that recognizes linkages among system elements
Attributes of context in which plans are developed and implemented	
4. Iterative review and continuous learning	Regularly review plans to address emerging issues and trigger important policy adjustments
5. Multistakeholder deliberation	Improve legitimacy, salience, and comprehensiveness of planning decisions with deliberation among parties to the plan, recognizing an "open impartiality" that accepts legitimacy and the importance of the views of others
6. Diversity of approaches	Implement a variety of alternative policies to gain knowledge about the most effective approaches
7. Decentralized decisionmaking	Improve flexibility and responsiveness by placing decisionmaking authority and responsibility at the lowest effective and accountable level of governance

SOURCES: Adapted from Swanson et al., 2007; 2010.

plans and planning processes, drawn from the literature (see Appendix C), and particularly for the processes, customized as necessary for any particular application. Drawing from Table A.14, for example, a plan that clearly articulates its objectives, goals, and priorities and includes assessments of the needed human, social, natural, and other types of capital would tend to rank highly for quality. A planning process that included meaningful engagement with a full range of stakeholders and decisionmakers would tend to rank highly for process quality.

We have also included *cost-effectiveness* as an indicator of the quality of the planning process. The cost of planning includes the financial resources required, along with the staff time and the time and attention demanded from the community. We envision the effectiveness of the planning process as a counterfactual: Would there have been a less resource-intensive process for achieving the same level of quality?

Relationship of Decision Support Indicators to Other Indicator Systems

To place these categories of indicators for urban climate risk management in perspective, we compare them to related systems of indicators in the literature. Table 3.5 compares our five categories of indicators to the categories of indicators developed by Arup (2014) for the Rockefeller 100 Resilient Cities initiative. Arup's framework, which derives from a systems-focused view, focuses, not surprisingly, on well-being and capacity to implement, both associated with the well-being of the city. However, the framework also includes categories related to finances and leadership that address transformational capacity. Table 3.6 compares our proposed categories of indicators

Table 3.5
Comparison of Proposed Categories of Indicators and Arup Indicators of Urban Resilience

Arup Indicators		RAND Proposed Indicators				
		Well-Being and Risk to Well-Being	Capacity to Implement	Capacity to Adapt and Transform	Quality of Planning Process	Quality of Plan
Health and Well-Being	Minimal human vulnerability	X				
	Livelihoods and employment					
	Safeguards to life and health	X				
Economy and Society	Identity and mutual support	X	X			
	Social stability and security	X	X			
	Financial resources		X	X		
Urban System and Services	Reduced physical exposure	X	X			
	Continuity of services	X	X			
	Reliable communications and mobility		X			
Leadership and Strategy	Leadership and management		X	X		
	Empowered stakeholders		X	X	X	
	Integrated planning				X	X

SOURCE: Arup, 2014, and the authors.

Table 3.6
Comparison of Proposed Indicators and Preston et al.'s (2011) Adaptation Planning Stage Criteria

Preston Criteria		RAND Proposed Indicators				
		Well-Being and Risk to Well-Being	Capacity to Implement	Capacity to Adapt and Transform	Quality of Planning Process	Quality of Plan
Goal-Setting	Articulation of goals					X
	Success criteria					X
Stock-Taking	Human capital		X			
	Social capital		X			
	Natural capital		X			
	Physical capital		X			
	Financial capital		X			
Decisionmaking	Stakeholder engagement				X	
	Assess climate drivers					X
	Assess nonclimate drivers					X
	Assess impacts, vulnerability, and risk					X
	Acknowledge uncertainties					X
	Options appraisal					X
	Exploit synergies					X
	Mainstreaming		X		X	X
Implement and Evaluate	Communication and outreach		X		X	X
	Define roles and responsibly		X			X
	Implementation		X			X
	Monitor, evaluate, and review	X				

SOURCE: Preston et al., 2011, and the authors.

to Preston et al.'s 2011 framework for evaluating adaptation planning, summarized in Table A.13. The planning evaluation framework, which derives from the decision-focused logical framework analysis, focuses on the quality of the plan, the capacity to implement the plan, and the quality of the process used to develop it. Neither the Arup (2014) nor Preston et al. (2011) indicators explicitly call out indicators that capture the capacity for transformation.

Similar comparisons could be made to other indicator frameworks reviewed in Appendix C. As Tables 3.5 and 3.6 show, the proposed framework is consistent with, inclusive of, and can draw usefully from related frameworks in the literature. However, our assessment also suggests that no existing framework fully addresses the requirements we have set forth for guiding support for decisionmaking and assessing the effectiveness of an iterative climate risk management framework for urban areas.

Example Application of the Proposed Indicator System

Given the importance of context, we consider the application of the proposed indicators system within a specific, hypothetical case of a coastal metropolis faced with major water supply, drainage, and wastewater infrastructure and land use decisions. These decisions are driven by economic development goals but also contingent on the current and projected tax base of the region, as well as external factors, including the rate of sea-level rise and risk of damage from major coastal storms and flooding.

The urban area described as Case 2 in Chapter One includes several counties and cities, each with their own land use and transportation plans. Several federal and state agencies have large equities in the region in the form of management of large tracts of land and leading responsibility for some major flood control works. No single regional authority coordinates planning and implementation of major public infrastructure, with the exception of a regional water management agency that has lead responsibility for some of the region's water infrastructure. Transportation, land use, and water management decisions in one county can have collateral impacts on drainage, flooding, groundwater levels, and saltwater intrusion in another. The effectiveness of changes in land use policy in one county may depend on the extent of coordination among jurisdictions within each of the counties. Decisionmakers and stakeholders seek to understand how alternative development strategies and implementation of individual projects will help achieve the community's flood protection, water reliability, environmental, and economic goals in the face of large uncertainties regarding sea-level rise, precipitation patterns, and land use changes.

Table 3.7 summarizes the categories of indicators that might be used for this example and approaches to data gathering that might enable their use. Indicator types that are new to the climate risk management literature are shown in bold. In the

Table 3.7
Categories of Indicators Applicable to Example

	Examples of Classes of Indicators	Approaches to Data-Gathering
Well-Being and Risk to Well-Being	• Regional economic development • Flood risk • Water quality • Ecosystem protection • Financial solvency of public utilities • Equity of benefits and costs	• Environmental monitoring and socioeconomic data for current conditions • **Projections of future risk from risk assessment simulations**
Capacity to Implement (Within Tier 1)	• Financing/funding available • Incentives and their alignment for key actors • Institution, managerial, and technical capabilities • Organizational leadership and culture • Monitoring and evaluation • Public support	• Economic assessment of incentives and adequacy of financial resources • Surveys of institutional culture and public support • Qualitative assessments of other factors • **Projections of future risk from risk assessment simulations**
Capacity to Adapt and Transform	• Incentives • Public support • Leadership • **Assessment of type and magnitude of Tiers 2 and 3 changes needed to achieve benefits (e.g., risk reduction, economic growth, water quality improvement)** • **Complexity of changes**	• Economic assessment of incentives and adequacy of financial resources • Surveys of institutional culture and public support • Qualitative assessments of other factors • **Vulnerability analysis contingent on changes in Tiers 2 and 3**
Quality of Plan	• Breadth of futures and options • Use of best available science • Analysis of trade-offs among goals • Alignment of finances and funding with plan implementation • **Extent to which plan is designed for learning and adaptation**	• Checklist of plan attributes • Stock-taking of resources • **Gap analysis comparing plan to risk analysis, especially key factors driving vulnerability and potential responses**
Quality of Planning	• Openness and transparency of public process • Technical credibility • Political legitimacy • Salience to area's perceived problems and challenges • Cost-effectiveness	• Surveys and focus groups with stakeholders and decisionmakers • Tally of the resources required for the process, compared with estimated resources for other approaches with similar quality and outputs

remainder of this section, we describe and give examples of these indicators (Tables 3.8 through 3.12) in the context of this example.

Indicators of Well-Being and Risk to Well-Being

Indicators of well-being play a crucial role in the evaluation of efforts toward urban climate risk management. To be most meaningful in a decision context, these indicators should bear some relationship to the goals that are driving regional decisionmaking. Well-being indicators like those in Table 3.8 can be used to assess the current state of the region, assess potential future states of the region under various strategies

Table 3.8
Potential Categories (by Goal) and Indicators Associated with Well-Being and Risk to Well-Being

Category	Potential Indicators
Regional economic development	• Gross regional product • Economic losses from service interruptions
Flood risk	• Expected annual damage reduction
Water quality	• Net change in location of salt water/fresh water interface relative to current base (and/or business-as-usual projections) • Net change in nutrient loadings and concentrations
Ecosystem protection	• Area protected
Financial solvency of public utilities	• Net change in property tax revenues • Annual operating costs of water-related, wastewater-related, and transportation-related public services • Capital costs of new infrastructure (water, drainage, wastewater, transportation) • Annualized life cycle costs • Cost-effectiveness of new infrastructure relative to achievement of other goals
Equity of costs and benefits	• Gap in costs and benefits between upper and lower 10 percent of affected populations

and scenarios, and organize monitoring activities to track progress over time. In our example, the region is seeking to improve various infrastructure systems to reduce flood risk, maintain reliable fresh water supplies, restore aquatic and wetlands ecosystems, and sustain economic development. As listed in Table 3.8, appropriate well-being indicators might therefore relate to mitigating flood risk, limiting cost and disruption of public services and economic activity associated with reducing flood vulnerability, minimizing the impact of saltwater intrusion on drinking water, and enhancing the functionality of aquatic ecosystems.

Many of the systems of indicators reviewed in Appendix C offer other similar examples of well-being indicators. Such indicators are a major focus of the Rockefeller indicators (Arup, 2014) for urban resilience (see Table 3.4), which include six types of such indicators in the categories of health and well-being, economy and society, and urban systems and services. The TAMD Track 2 (Table C.6) outcome indicators also address well-being, focusing on such factors as vulnerability to climate-related hazards, changes in poverty, and other development indicators (Brooks et al., 2013). Similarly, the outcome criterion in the Baker et al. (2012) framework addresses well-being, with a focus on such attributes as water quality, incidents of extreme heat, and risk of wildfires. Well-being indicators can also be drawn from assessments of vulnerability in the adaptation literature.

Simulation models can be used in risk management analysis to project some of the future values of the indicators, contingent on proposed plans and various uncertainties. These projections play a central role in evaluating and comparing the potential consequences of alternative urban climate risk management strategies. In our proposed example, several independent models relating precipitation patterns and sea level to surface- and ground-water management and water quality constitute an integrated simulation model that can be used to understand current and projected future conditions under different strategies or plans and scenarios that capture key uncertainties.

Indicators of Capacity to Implement

Some of the indicators systems in Appendix C offer indicators relevant to capacity to implement, although they are not called out as such. For instance, the Rockefeller indicators for urban resilience (Arup, 2014), shown in Table C.8, include availability of financial resources and contingency funds in the category of economy and society and effective leadership and management and integrated development planning under the category of leadership and strategy. Similarly, the Preston et al. (2011) framework provides such indicators in its stock-taking, decisionmaking, and implementation and evaluation categories in Table C.13. *Implementation* is broadly construed in this framework to include communications and outreach, definition of roles and responsibilities, monitoring, evaluation, and review. In the literature on barriers to adaptation, for instance, the Ford and King (2013) framework for adaptation readiness could also prove useful here. Common determinants of the capacity to implement plans include available financing mechanisms, funding, information, infrastructure, and other resources.

Drawing on these sources, we have identified indicators of capacity to implement that would be applicable to our example, as shown in Table 3.9. Current operations and planning reside entirely within Tier 1. Hence, the region might measure its current capacity to implement by qualitative assessments of such factors as the availability of appropriate information, institutional capability, and institutional alignment; economic assessments of the available financial resources and the incentives on various parties; and surveys of institutional culture and public opinion.

Indicators of Capacity to Adapt and Transform

An assessment of capacity to implement could lead to a finding that the region's capacity to adapt and transform is low because of insufficient financial resources or because regional actors may lack enthusiasm for making adjustments or overhauling existing governance structures in ways that would increase the region's overall effectiveness in infrastructure planning and construction. Current operations could be limited in the level of risk that can be reduced. At the Tier 1 level, most of the region's cities in our example lack the necessary financial resources, technical expertise, and planning culture to develop and implement comprehensive flood control and land use plans that could sustain long-term risk reduction.

Table 3.9
Potential Indicators Associated with Capacity to Implement

Category	Potential Indicators
Available financing and funding	• Sufficiency of dedicated revenue streams • Status of relevant public budgets (surplus or deficit; sustainability of funding from one administration to the next) • Size of contingency funds relative to total operational budget
Incentives and their alignment for key actors	• Willingness to negotiate (yes or no) • Willingness to compromise (yes or no)
Institutional, managerial, and technical capabilities	• Availability of appropriate technical information, as determined by a technical peer review process • Capacity to manage complex technical contracts (yes or no)
Organizational leadership and culture	• Single point of decisionmaking on storm water management (yes or no) • Mechanism for coordinating decisions on storm water, land use, and other key aspects of urban services (yes or no) • Stability in leadership and management (average tenure of top leaders and managers)
Monitoring and evaluation	• Measures that track progress on implementation (yes or no) • Procedures for formal evaluation by third-party evaluators (yes or no)
Public support	• Percentage of support by voters as expressed in ballot initiatives, referenda, or public opinion polls

Meeting the region's goals for flood risk reduction and water quality could, therefore, require changes in Tier 2 and perhaps Tier 3. Tier 2 changes might involve multi-municipal solutions and intergovernmental agreements in which communities join together to jointly develop and fund common infrastructure projects and coordinate regulations. In addition, communities could join together to support joint investments in technical and planning expertise. Tier 3 changes might involve developing a new regional authority that would have the ability to levy fees, invest in joint projects, provide technical assistance, and coordinate local storm water ordinances. (A Helm and Sprinz scale in Figure 3.2 might suggest that such Tier 2 changes would enable the region to achieve risk reduction closer to a collective optimum than in Tier 1, and that Tier 3 changes could move the region closer still.)

An assessment of the current capability to transform in Tier 2 might examine the incentives for various communities that might cause them to view voluntary agreements favorably or unfavorably, the views of communities on such agreements, the complexity of such agreements, any local tradition with such institutions, and the presence or absence of leaders eager to make such changes. An assessment of the capacity to transform to Tier 3 would include a review of existing legal authority to create such an organization. Such an assessment could also survey the community for their views on creating such an organization, evaluate the incentives on various communities that might cause them to view such an organization favorably or unfavorably, and note any tradition of such agencies in the region.

Once specific changes of interest have been identified in Tiers 2 and 3, measures of capacity to make these transformations are needed as identified in Table 3.10. These measures could be constructed using economic analysis of the incentives on various actors favoring or disfavoring the transformation and of the various financial implications, surveys of institutional culture and public support, and qualitative assessments of such factors as leadership, the consistency of the changes with the political and social culture of the region, and the complexity of the envisioned changes.

Indicators of the Quality of Plan

Overall, evaluating plan quality can focus on checklists and gap analysis. The former help to measure whether the plan includes all of the important components. The latter can compare the actions, futures, and perspectives considered in the plan to those that might emerge as potentially important in an unconstrained risk management analysis. For instance, a gap analysis can address the potential for maladaptation, especially over the long term, by comparing the key drivers of vulnerability and the responses to them that are included in the plan to those that emerge (or might have reasonably emerged) from the quantitative risk management analysis.

Table 3.10
Potential Indicators Associated with Capacity to Adapt and Transform

Category	Potential Indicators
Incentives of actors to favor transformation	• Percentage of officials, community, and interest group thought leaders who believe that they are incentivized to support proposed changes, based on third-party survey or focus groups
Public support	• Percentage of support by voters, as expressed in ballot initiatives, referenda, or public opinion polls • Percentage of voters who perceive that change is possible, as measured in credible polling or focus groups
Leadership	• Existence of a clearly identified champion in a position of authority to lead the transformation effort (yes or no) • Percentage of officials, community, and interest group thought leaders who believe that their region's institutional and social culture can support change, based on third-party survey or focus groups
Assessment of type and magnitude of Tiers 2 and 3 changes	• Number of regulatory changes required for each change • Number of statutory changes required for each change • Number of new governance structures required • Policy analysis of potential changes, including degree of difficulty (yes or no)
Complexity of changes	• Number of public agencies with equities in the decision

As shown in Table 3.11 for our example, the quality of the region's plan might be measured by the extent to which it reflects the full range of the community's goals, considers a wide range of options for reaching those goals, examines the efficacy of those options over a wide range of futures, includes attributes of adaptive plans, includes consideration of the benefits and costs in the three tiers, and includes options for changes in those tiers.

Quality of the Planning Process

Indicators describing the quality of the planning process are intended to capture the inclusiveness of the process, the extent to which it conforms to expected norms, and the extent to which it is seen as legitimate. Such attributes are important for the effectiveness of the plan and also represent attributes of intrinsic value. Preston et al. (2011) offer three types of indicators for the planning process in the categories of decisionmaking and implement and evaluate. The Rockefeller indicators (Arup, 2014) for urban resilience suggest two such indicators, both in the category of leadership and strategy.

In our example, the quality of the planning process would depend on the extent to which all of the relevant stakeholders had an opportunity to engage with the process and the extent to which they found it legitimate, credible, and salient. Evaluating the quality of planning can largely focus on survey methods that solicit people's views of the process. It is important, however, to ensure that such surveys are broadly inclusive of the community, in particular of those whose were not represented in the process but whose inclusion might have brought a significantly different viewpoint to the process. Table 3.12 is suggestive of the types of indicators that could represent plan quality.

Table 3.11
Potential Indicators Associated with Quality of Plan

Category	Potential Indicators
Openness and transparency of public processes	• Percentage of community and interest group thought leaders satisfied with the process, based on third-party evaluation survey
Breadth of options considered	• Percentage of officials, community, and interest group thought leaders satisfied with the breadth of options considered, based on third-party evaluation survey
Use of best available science	• Successful completion of technical peer review process on analysis underlying plan (yes or no)
Analysis of trade-offs among goals across scenarios (robustness)	• Visualization of trade-offs across goals (yes or no) • Estimated success or failure rate of plan across wide range of futures at specified time intervals, as generated by the RDM process
Alignment of finances and funding with plan implementation	• Existence of credible financing and funding plan to cover full or staged implementation of plan (yes or no)
Extent to which plan is designed for learning and adaptation	• Adaptive planning process established (yes or no) and feasible options likely to be available at the time of plan update (yes or no)

Table 3.12
Potential Indicators Associated with Quality of Planning Process

Category	Potential Indicators
Technical credibility	• Weight of responses from technical peer reviewers to the analytical products generated in the planning process
Political legitimacy	• Percentage of officials, community, and interest group leaders who believe that the process was inclusive, fair, and equitable, based on third-party evaluation survey
Salience to area's perceived problems and challenges	• Public statements made by lead decisionmakers on the timeliness of the plan
Cost-effectiveness	• Tally of resources required, compared to alternative approaches

Role of Simulation and Logic Models

Simulation models of physical processes are one of the tools that can be used in Tier 1 to relate actions to consequences. Using principles of hydrology and physics, a groundwater flow model, for example, might help to explain how sea-level rise could affect groundwater levels and flooding in an urban coastal region. Another example could be a transportation demand model that helps regional planning authorities relate changes in land use and economic transformation in a region to likely changes in demand for public transit, rail, and roads. These mathematical simulation models typically embody assumptions about current policy and operating rules, hence their suitability for Tier 1 analysis.

Logic models are another means of connecting actions to consequences, and thus they gather and array information about the effectiveness of organizational (or multi-organizational) strategy and performance based on inputs, implementing processes, and indicators of progress (Greenfield et al., 2006). Logic models can be used to supplement simulation models by capturing elements of implementation that fall outside of the mechanistic or statistical processes that lend themselves to mathematical modeling. Figure 3.4 offers an example of the details typically included in a logic model that disaggregates inputs and sequences actions, outputs, and outcomes. This particular example applies to Tier 1 "business-as-usual" processes and was developed to aid in evaluating the effectiveness of RAND's decision support interventions in three urban areas—each seeking in different ways to understand and evaluate potential responses to climate change, as manifested in major policy and investment decisions (Melissa Finucane, personal communication, 2016).

Figure 3.4
Example of Logic Model For Tier 1 Processes

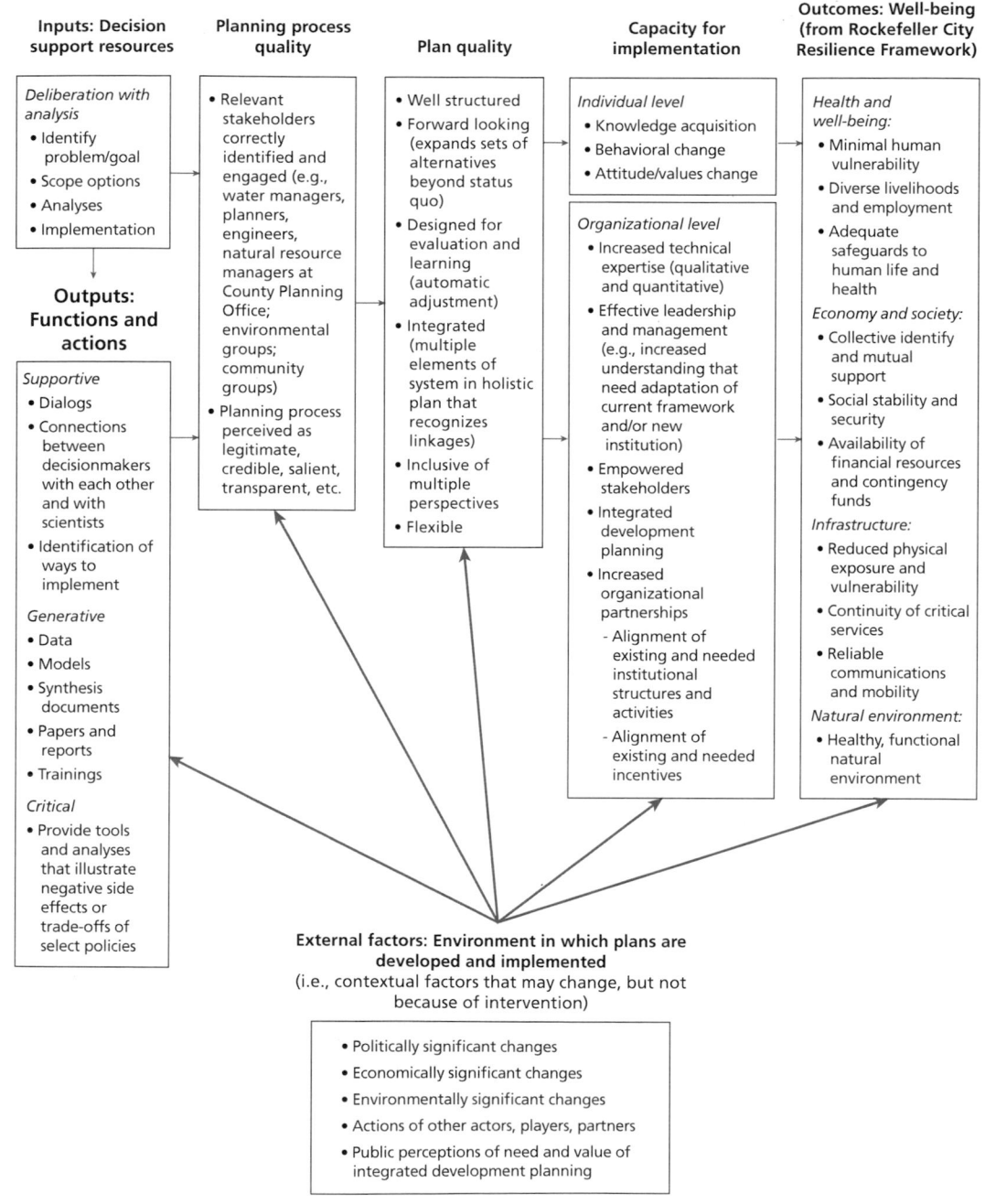

SOURCE: Melissa Finucane, RAND Corporation, 2016.
RAND RR1144-3.4

Summary of Findings

The process of choosing specific indicators in the context of urban responses to climate change must always be grounded in the particular values, goals, and ambitions of the region itself. For this reason, we have focused on categories of indicators from which specific indicators could be selected, including those in Tables 3.8 through 3.12 accompanying the example. While the literature on indicators to inform projects and initiatives related to resiliency, adaptation, and mitigation is growing by leaps and bounds, use of these indicators is typically not placed in the context of shaping the planning process and framing the choices for decisionmakers. Our intent in this chapter has been to demonstrate what such a mapping might look like and to go one step further by demonstrating how different categories of indicators relate to the potential for transformation of governance and decisionmaking to rise to the need for more effective responses to climate change in the context of the larger set of issues facing urban areas.

CHAPTER FOUR

Discussion

An increasing number of urban areas in the United States and throughout the world are engaging in mitigation and adaptation activities—whether tied directly to climate change or as a means to build resilience against extreme weather, natural disasters, and other stresses and disruptions (see Appendix D for an overview). Activities extend from public awareness to community resilience checklists. In a few places, major problem-focused infrastructure planning is under way to deal with coastal planning or aging storm water systems. California, with its statutory limits on GHG emissions, also has a law to drive changes in land use and transportation planning in major urban areas toward lower net emissions.

The shaping and setting of goals, plans, implementation strategies, and operating rules for these response actions depend on how decisionmakers frame the overarching climate change threat. For example, the need for mitigation or adaptation could be posed as a technical constraint on business as usual that could be loosened by new technologies of various kinds and the use of "big data" to identify opportunities and threats to well-being. Alternatively, the challenge could be more appropriately framed as one of managing risk, with near-term costs and deeply uncertain long-term benefits. Going one step farther, the threat of climate change could be viewed as posing a fundamental challenge to urban governance as we know it. In this framing, old models of stovepiped bureaucracies and public works and service delivery are too rigid and constricting to meet the demands of long-term, cross-sectoral, highly uncertain, and potentially high-impact change.

In Chapter Two, we posited that risk governance, a more general form of risk management, is an appropriate framework because of its capacity to encompass a multiple-actor, decision-centric perspective and its recognition that climate change responses are more than technical in nature. While many climate change response choices—like improvements in energy efficiency, expanded public transportation, or more progressive building codes in flood-prone areas—fall well within business-as-usual operations at the city level, other alternatives are more challenging because of their larger regional impacts; their overlap across public agency jurisdictions; and their requirements for new policies, regulations, and laws in some cases. New or recapitalized infrastructure, land use changes, and multijurisdictional financing for regional storm water management are examples. To implement risk governance under such conditions, we add to the framework three important enhancements: decision support under conditions of deep uncertainty, iterative risk management and learning, and the idea of "tiers of transformation" to capture the need for varying degrees of transformation in governance that may be needed to reduce risk to acceptable levels.

Having an appropriate conceptual framework, however, is not enough to guide action. The framework needs to be operationalized within a decisionmaking process, preferably one that makes use of transparent, participatory planning and objective analysis to support deliberations. Further, indicators are essential as part of decision support to accurately capture progress toward the desired outcome of risk reduction, as well as the indicators to capture capacity building for transformation and implementation outside the norms of business as usual. Many systems of indicators have been proposed over the last decade or so, as we describe in Chapter Three and Appendix C, that have some relevance to how cities, regions, and nations build capacity and respond to climate change, resilience, or whatever their construct might be for thinking about planning for long-term change. We found that no single existing system of indicators captured the full dimensions of our proposed conceptual framework, including not only risk management but also considerations of governance and the quality of the planning and decision support processes themselves. Instead, we have chosen to synthesize a system of indicators, drawing from indicators proposed by others and adding new indicators as needed, as summarized in Table 3.4 and the example in Table 3.7.

Steps Toward Applying the Framework

Decision processes at any level of government can prove messy. Within U.S. cities, major infrastructure decisions tend to be made within "line" departments, such as public works, transportation, water and sewers, environmental protection, and housing. As a matter of course, these departments tend to follow their own planning and procurement procedures. While their representatives may gather regularly to compare notes and coordinate on specific individual actions, they tend not to conduct major planning processes with one another. This is the reason why the Rockefeller Foundation, for example, is supporting resilience officers through its 100 Resilient Cities program. These resilience officers are intended to help break down stovepipes of governance and help forge solutions that cut across traditional lines of responsibility within cities. But the governance issue goes deeper. As one example, not only will cities need to find new ways to respond, but cities will also need to find ways to work cooperatively with their surrounding suburban and exurban areas to implement effective, long-term solutions. This is even more difficult because of the variance among jurisdictions in economic well-being, taxation, income disparities, and other socioeconomic factors.

The decision support literature, along with the experience of the authors and our colleagues, suggests several important lessons for implementing the risk governance framework under such conditions. The first step is to find an appropriate forum for deliberative decisionmaking or, if none exists, to create one. In December 2005, following Hurricane Katrina, the State of Louisiana formed the Coastal Protection and Restoration Authority (CPRA) to cut across its multiple state government departments with responsibilities in coastal Louisiana (CPRA, 2015). CPRA also was intended to address the needs of the entire coastal region, not just New Orleans, Port Charles, and other urban areas along the coast. Forming a new governance structure for regional decisionmaking can prove difficult in current American politics. The challenges of climate change may require that such changes become more frequent when existing governance arrangements fail to demonstrate the capacity to reduce risks in socially and economically acceptable ways.

A second step toward implementation, closely tied to the first, is finding visionary leaders who can steer transparent and deliberative processes toward a successful conclusion by embracing democratic processes within the context of a "deliberation with analysis" approach. That is, strong leadership can set the tone for elevating objective analysis to its proper place within participatory planning and decisionmaking—not as a means to justify a predetermined decision, but rather as a means of opening the way to consideration of a range of alternatives and trade-offs among them in terms that are clear and meaningful to the public and their representatives.

Finally, a third critical step is the harnessing of both community thought leaders and technical expertise to play leading roles in the deliberative process from the very start. Effective risk communications should begin at the very earliest stages in a planning process. As the work of Kahan et al. (2014) and others has shown, credible voices coming from different perspectives within the community are critical to building understanding among the broader public. Public support in turn lends credibility to the planning and decisionmaking processes. Locally based technical leadership adds further connective tissue between the public and the process. Scientists and engineers from local and regional institutions tend to have strong understanding of local conditions and a heightened appreciation of the interplay of science and social issues within their areas by virtue of being part of the affected communities.

Measuring Effectiveness of Processes and Outcomes

Chapter Three and Appendix C provide an extensive discussion of the many types of indicators proposed for use in adaptation, resilience, and other related activities in urban areas. In our proposed system of indicators, indicators of well-being represent the top-level outcomes that policies and strategies intend to achieve. Practically speaking, changes in these indicators may take years or even longer to materialize as planning and implementation of response measures take shape and become reality. While to some extent it is possible to estimate the future consequences of current actions using some combination of simulation models and qualitative judgments, such estimates are unavoidably and often irreducibly uncertain.[1] For this reason, it also proves necessary to complement indicators focused on outcomes with indicators focused on plan quality and process. In addition, these latter indicators can incorporate issues of justice, fairness, rights, and inclusivity that are poorly captured by consequentialist indicators focused only on outcomes (Jones et al., 2014).

Collecting the data to populate the indicators across the five major categories listed in Table 3.6 will require an evaluator who largely sits outside of the deliberative process itself to administer surveys, analyze results, and direct the collection of data and information from other sources. To be effective in measuring change—and progress—over time, an evaluator will need to be present from the creation of the process to capture baseline attitudes of participants in the planning process and identify data needs along the way.

In contrast with indicators devised primarily as "report cards" or to support a "one-off" development project, we advocate for a system of indicators embedded in long-term planning and decisionmaking processes as a means of ensuring continuity and focus. Processes like Louisiana's Comprehensive Master Plan for a Sustainable Coast, the Comprehensive Everglades Restoration Plan, and the Metropolitan Water District's Integrated Resources Plan have been designed to provide stability and continuity in planning for investment and management of large and lasting regional infrastructure. In these examples, indicators of progress are essential drivers of the processes themselves, giving resource managers and their agencies a strong motivation for maintaining the systems necessary to collect, process, and actively use the indicators with decisionmakers and the public.

Moving from Theory to Practice

Many systems of indicators have been devised for use by cities; relatively few are maintained and kept timely and fresh beyond their first few years. Even fewer systems of indicators have been tested under varying conditions and evaluated for their effectiveness. In the course of conducting this review, we have come to appreciate the need for a generalizable set of indicators that urban areas new to the long-term, integrated planning process could adapt for their own purposes. To reach that goal, however, requires further development, field testing, and evaluation.

---

[1]   Even in hindsight it is often not possible to definitively make the connection between action and consequence. To the extent that one can observe only one actual realization of what might have been many historical paths, it proves difficult to separate contingency from determined outcome (see, for instance, March et al., 1991).

RAND has launched three such field studies in the Pittsburgh region, southeast Florida, and the Bay Area megaregion of northern California. The engagement with each area is following a "deliberation with analysis" process. In each, RAND is working with a regional convening forum for deliberations: the Allegheny County Executive in the first case, the Southeast Florida Regional Climate Change Compact counties in the second, and in the third case, three MPOs: the Sacramento Association of Governments, the MTC, and the San Joaquin Association of Governments. As described in the example at the end of Chapter Three, indicators are being collected to support evaluation of the planning process, the tracking of outcomes, and assessment of capacities for transformation and implementation. This approach will be useful for evaluation of the studies and will also enable iterative learning within the decision support processes. Evaluation began at the initiation of each engagement and will carry through to the end of RAND's engagement, at a minimum.

RAND has been working in close collaboration with regional leaders and other key players to scope, design, and implement an interactive planning process that fits the region's needs. The first several meetings were used to clearly articulate the problem to be solved, the goals to be achieved in a solution, and the options to address the problem. In this crucial step of decision structuring, decisionmakers discuss and select the key indicators that represent their multiple planning goals.

RAND then turned to building a decision support tool (known as the Planning Tool), customized to each application, to organize information about the system of interest, policy or investment options, and indicators of interest when assessing the options and trade-offs among goals. At this stage of the process in each of the field studies, indicators associated with capacity to implement and transform have been identified.

The next set of meetings are of a more technical nature, as data and models are gathered to help analyze a future without action and the potential benefits and costs of various options. Output from these models is incorporated into the customized Planning Tool, which includes visualization software to help interpret results. The final set of meetings will be interactive sessions with decisionmakers (and stakeholders when feasible) to explore the range of possible outcomes under different options and scenarios of climate change and other key uncertainty factors affecting future benefits and costs. Each study has been designed to run for about 12 months, although arrangements may be made to extend one or more of these studies further. In these efforts, RAND has taken the lead on facilitation and analysis in the spirit of capacity building. In future efforts, we expect local partners to carry on these processes on their own.

Simulation and visualization tools are critical elements of the deliberative process. Each study is using simulation models drawn from existing models in the region and customized for their specific context. RAND's previous analytic facilitation efforts in Louisiana (Groves et al., 2014), the Colorado River Basin (Groves, Fischbach, et al., 2013), and elsewhere used detailed simulation models trusted by the stakeholders to relate actions (policy choices) to consequences (change in risk). They also used visualization tools to help the stakeholders envision vulnerabilities and trade-offs, engage in an iterative process of answering "what if" questions with the data, and deliberate with one another over decision options. We have taken a similar approach with these three field studies.

In addition to addressing priority decisionmaking needs in each of the three urban areas, these field studies also provide a means of testing the validity and utility of the decision support framework and system of indicators proposed in this report. We will publish the results of our assessment following the completion of all three studies.

Concluding Remarks

In this review, we characterized the current state of the art of applying indicators to responses to changing climate in urban areas and placing those responses in a context of risk governance and management, long-term planning, and decisionmaking under uncertainty. Cities in the United States and throughout the world are in a period of rapid change as a consequence of many forces, with climate change being only one of those. As such, planning for climate change—whether oriented toward mitigation, adaptation, or both—is occurring in a complex governance and political environment with intense competition for resources to address a range of societal needs. Thoughtful, visionary leadership is essential for achieving acceptable levels of distributional equity among urban residents and striking an appropriate balance of near-term and longer-term needs. We have no process to prescribe for producing such leadership, but we do hypothesize that the framework and indicators discussed in this report could be helpful to those who choose or are chosen to lead.

Literature Review of Key Terms

In this appendix, we summarize the many definitions found in the literature on terms that we define and use in the report. Our preferred definitions are in Tables 1.1 and 1.2 in Chapter One.

- **Adaptation**
  - "The process of adjustment to actual or expected climate and its effects. In human systems, adaptation seeks to moderate or avoid harm or exploit beneficial opportunities. In some natural systems, human intervention may facilitate adjustment to expected climate and its effects" (IPCC, 2014c, p. 118).
  - "Adaptation refers to adjustments in ecological, social, or economic systems in response to actual or expected climatic stimuli and their effects or impacts. It refers to changes in processes, practices, and structures to moderate potential damages or to benefit from opportunities associated with climate change" (UNFCCC, 2014).
  - "Adapting today is about reducing vulnerabilities to emerging or future impacts that could become seriously disruptive if we do not begin to identify response options now; in other words, adaptation today is essentially a risk management strategy" (National Research Council, 2010, p. 1).
  - "Adaptation consists of actions undertaken to reduce the adverse consequences of climate change, as well as to harness any beneficial opportunities. Adaptation actions aim to reduce the impacts of climate stresses on human and natural systems" (National Climate Change Adaptation Research Facility, undated).

- **Capacity to implement**
  - Having the resources and ability to engage in "a specified set of activities designed to put into practice an activity or program of known dimensions" (Fixsen et al., 2005). This definition is derived from the quoted definition of the word *implementation* and the definition of *capacity*.
  - The homepage of the National Implementation Research Network (2016) states: "According to this definition, implementation processes are purposeful and are described in sufficient detail such that independent observers can

detect the presence and strength of the 'specific set of activities' related to implementation. In addition, the activity or program being implemented is described in sufficient detail so that independent observers can detect its presence and strength."

- **Capacity to transform**
  - Having the resources and ability to produce a change in "the structural contexts and factors that determine [one's] frame of reference" (Pahl-Wostl, 2009, p. 359)
  - Having the resources and ability to produce "a change in the fundamental attributes of natural and human systems" (IPCC, 2014a). This definition is derived from the quoted definition of the word *implementation* and the definition of *capacity*.

- **Framework**
  - "The basic structure of something: a set of ideas or facts that provide support for something" (Merriam-Webster, undated)
  - "A conceptual framework is defined as a network or a 'plane' of linked concepts" (Jabareen, 2009, p. 49).
  - "The conceptual framework of your study—the system of concepts, assumptions, expectations, beliefs, and theories that supports and informs your research—is a key part of your design" (Maxwell, 2013, p. 39). Miles and Huberman (1994) defined a conceptual framework as a visual or written product, one that "explains, either graphically or in narrative form, the main things to be studied—the key factors, concepts, or variables—and the presumed relationships among them" (p. 18).
  - A logical framework is a "[m]anagement tool used to improve the design of interventions, most often at the project level. It involves identifying strategic elements (inputs, outputs, outcomes, impact) and their causal relationships, indicators, and the assumptions or risks that may influence success and failure. It thus facilitates planning, execution and evaluation of a development intervention. Related term: results based management" (OECD, 2002).

- **Impact**
  - "The fundamental intended or unintended change occurring in organizations, communities or systems as a result of program activities within 7 to 10 years" (Kellogg, 2004)
  - "A results or effect that is caused by or attributable to a project or program. Impact is often used to refer to higher level effects of a program that occur in the medium or long term, and can be intended or unintended and positive or negative" (USAID, 2009).

- **Indicator**
  - "Measurable representation of the condition or status of operations, management or conditions" (ISO, 2015)
  - "Quantitative or qualitative factor or variable that provides a simple and reliable means to measure achievement, to reflect the changes connected to an intervention, or to help assess the performance of a development actor" (OECD, 2002)
  - "Quantitative or qualitative variable that provides reliable means to measure a particular phenomenon or attribute" (USAID, 2009)
  - "A specific, observable, and measurable characteristic or change that shows the progress a program is making toward achieving a specified outcome" (CDC, 2012)

- **Input**
  - "The human, financial, organizational, and community resources a program has available to direct toward doing the work" (Kellogg, 2004)
  - "Resources provided for program implementation. Examples are money, staff, time, facilities, equipment, etc." (USAID, 2009).

- **Iterative risk management**
  - "Refers to an ongoing process of identifying risks and response options, advancing a portfolio of actions that emphasize risk reduction and are robust across a range of possible futures, and revising responses over time to take advantage of new knowledge. Iterative risk management strategies must be durable enough to promote sustained progress and long-term investments, yet sufficiently flexible to take advantage of improvements in knowledge, tools, and technologies over time" (NRC, 2011, pp. 1–2).
  - "An established approach that uses a monitoring, research, evaluation and learning process cycle to improve management strategies. The approach is very flexible and can be applied to projects or sector analysis" (Weadapt.org, 2014).

- **Measure**
  - "A measure is a value that is quantified against a standard. A project implementer wants to know that the total urban green space being developed is 250 acres in size. 'Acres' is the standard, and all can see and agree on the size involved" (Viggh et al., 2015, p. 73).

- **Metric**
  - "A calculated or composite measure or quantitative indicator based upon two or more indicators or measures. Indicators help put a variable in relation to one or more other dimensions" (Viggh et al., 2015, p. 73).

- **Mitigation**
  - Mitigation of climate change: "A human intervention to reduce the sources or enhance the sinks of greenhouse gases" (IPCC, 2014a)
  - Mitigation of disaster risk and disaster: "The lessening of the potential adverse impacts of physical hazards (including those that are human-induced) through actions that reduce hazard, exposure, and vulnerability" (IPCC, 2014a)
  - "Climate change mitigation refers to efforts to reduce or prevent emission of greenhouse gases" (United Nations Environment Programme [UNEP], 2015).

- **Outcome**
  - "Specific changes in program participants' behavior, knowledge, skills, status and level of functioning. Short-term outcomes should be attainable within 1 to 3 years, while longer-term outcomes should be achievable within a 4 to 6 year timeframe. The logical progression from short-term to long-term outcomes should be reflected in impact occurring within about 7 to 10 years" (Kellogg, 2004).
  - "A results or effect that is caused by or attributable to the project, program or policy. Outcome is often used to refer to more immediate and intended effects. Related terms: result, effect" (USAID, 2009).
  - "The likely or achieved short-term and medium-term effects of an intervention's outputs. Related terms: result, outputs, impacts, effect" (OECD, 2002).

- **Output**
  - "The direct products of program activities [which] may include types, levels and targets of services to be delivered by the program" (Kellogg, 2004)
  - "The products, goods, and services which result from an intervention" (USAID, 2009)
  - "The products, capital goods and services which result from a development intervention; may also include changes resulting from the intervention which are relevant to the achievement of outcomes" (OECD, 2002)

- **Process**
  - "Set of interrelated or interacting activities which transforms inputs into outputs" (ISO, 2015)
  - "The programmed, sequenced set of things actually done to carry out a program or project" (USAID, 2009)

- **Quality of plan**
  - Degree to which a plan is able to achieve its intended impact

- **Resilience**
  - "The capacity of social, economic, and environmental systems to cope with a hazardous event or trend or disturbance, responding or reorganizing in ways that maintain their essential function, identity, and structure, while also maintaining the capacity for adaptation, learning, and transformation" (IPCC, 2014a)
  - "As defined in this report, resilience is the ability to prepare and plan for, absorb, recover from, and more successfully adapt to adverse events" (National Research Council, 2012, p. 1).
  - "Resilience is the capacity of a social-ecological system to absorb or withstand perturbations and other stressors such that the system remains within the same regime, essentially maintaining its structure and functions. It describes the degree to which the system is capable of self-organization, learning and adaptation (Holling 1973, Gunderson & Holling 2002, Walker et al. 2004)" (Resalliance.org, 2015).

- **Well-being**
  - Well-being is "the balance point between an individual's resource pool and the challenges faced" (Dodge et al., 2012).
  - "Well-being is most usefully thought of as the dynamic process that gives people a sense of how their lives are going, through the interaction between their circumstances, activities and psychological resources or 'mental capital'" (National Accounts of Well-Being, undated).
  - "Well-being refers to being well in general rather than within any specific area of life. In keeping with this relatively common usage, we define consummate well-being as an overall evaluation of an individual's life in all its aspects" (Diener, 2009, p. 9).

Survey of Frameworks for Urban Responses to Climate Change

In this review of relevant conceptual frameworks, we consider risk governance, vulnerability, barriers, resilience, the precautionary principle, control theory, and game theory, noting that these are often complementary and not mutually exclusive concepts. These choices emerged from a broad survey of the literature and represent the frameworks most commonly used. We also identified several less-used frameworks that provide additional perspectives. Each approach, not surprisingly, suggests an alternative framing of the challenge of an urban response to climate change. Note that this appendix describes these frameworks using the language that their creators or third-party observers use to describe them and makes no attempt to ensure that that language is consistent with the definitions in Tables 1.1 and 1.2 applied in this report.

As shown by the axes in Figure B.1, we compare these alternative approaches according to two key attributes particularly salient in framing risk governance: (1) the extent to which the approach takes a view of the system versus the decision actor and (2) the extent to which the approach focuses on one or more actors. A system view seeks to understand how the various components of a system interact and influence the overall behavior of the system and that of its individual components. In contrast, a decision-actor view may also consider the system as a whole but privileges certain elements within the system as decisionmakers who can make choices, albeit within constraints. The consequences of those choices depend on the interacting responses of other parts of the system. Within the decision-actor view, one can focus on single actors or on the independently made choices of several actors. These attributes—system versus decision-actor and single or multiactor—prove important for understanding the utility of indicators, which aim to inform the choices of one or more actors operating within a complex, interacting system. For reasons discussed later in this appendix, we argue here that risk governance provides the most useful and encompassing framework for addressing the challenge of urban responses to climate change and is largely inclusive of the others.

Figure B.1
Comparison of Conceptual Frameworks

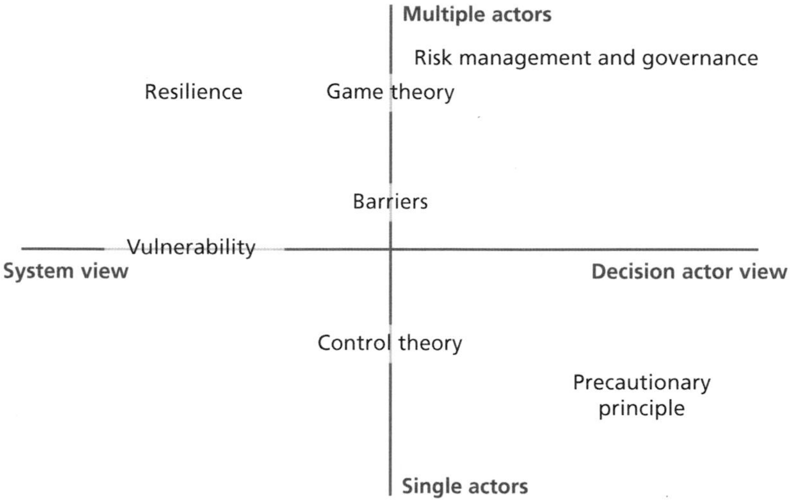

```
                                    │ Multiple actors
                                    │
                                    │  Risk management and governance
                                    │
            Resilience        Game │theory
                                    │
                                    │
                                    │
                                    │
                              Barr│iers
                                    │
─────────── Vulnerability ──────────┼─────────────────────────────────
  System view                       │           Decision actor view
                                    │
                                    │
                          Contro│l theory
                                    │
                                    │            Precautionary
                                    │              principle
                                    │
                                    │
                                    │ Single actors
```

RAND *RR1144-B.1*

Vulnerability

The concept of vulnerability plays a central role in the climate change adaptation and disaster risk management literatures (IPCC, 2012). Similar physical events, such as storms or droughts, can lead to very dissimilar effects on different people and places. For instance, hurricanes or tropical cyclones of similar intensity often cause far greater loss of life in poor countries than they do in rich countries, while rich countries often suffer larger economic losses. Defined as "the propensity or predisposition to be adversely affected" (IPCC, 2014c, p. 128), the concept of vulnerability helps to capture the sources of such differences.

The literature, particularly in disaster risk management, has increasingly emphasized the concept of vulnerability. The early disaster risk literature focused on risk due to physical events, such as floods, winds, or drought. However, vulnerability also encourages a broader understanding of socioeconomic factors that may be significant determinants of the level of risk. Such factors may include wealth and income, population growth, patterns and rates of urban development, socioeconomic inequalities, social cohesiveness, financial resources, the extent and attributes of infrastructure, gender, ethnic diversity, and governance.

Assessing vulnerability generally constitutes an important and early step in most climate change adaptation planning. For example, the World Bank's Urban Risk Assessment framework (Dickson et al., 2012) suggests that cities first evaluate the physical climate hazards they face (Hazard Impact Assessment) and then evaluate the vulnerability of various population groups, economic sectors, and geographic regions. Many of the indicator frameworks considered in Appendix C include assessments of vulnerability.[1]

Many frameworks exist in the literature for defining and measuring vulnerability. Of particular relevance, there exist indicators of social vulnerability that aim to identify the combinations of characteristics—such as income, ethnic diversity, social cohesion, and age—most closely correlated with adverse effects from physical hazards in any particular community. For instance, Figure B.2 illustrates the results of one such index of social vulnerability for the entire United States (Hazards and Vulnerability Research Institute [HVRI], 2013). This index includes over 30 measures that research

Figure B.2
Social Vulnerability Index

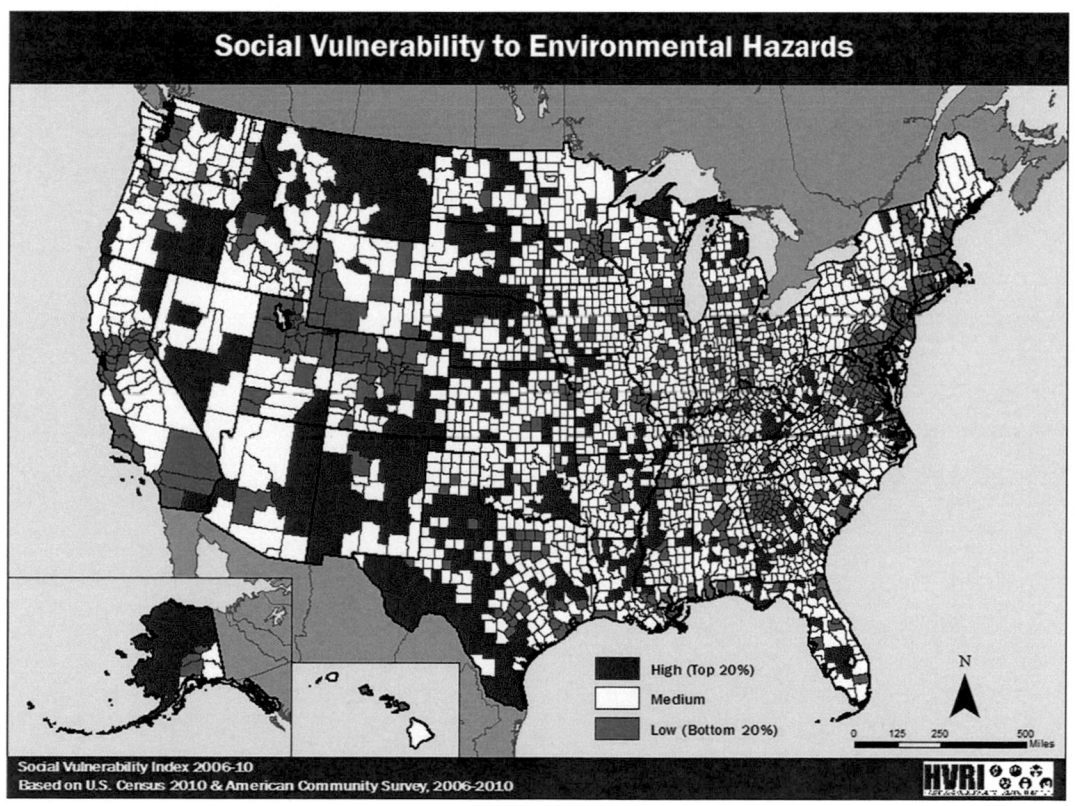

SOURCE: HVRI, University of South Carolina, 2013. Used with permission.
RAND RR1144-B.2

---

[1]   These include the assessment of the information base in the Queensland framework (Table C.4), one of the TAMD indicators (Table C.6), and the Rockefeller indicators of social well-being (Table C.8).

has shown to be most correlated with vulnerability[2] and suggests that many regions in the mountain West and along the Atlantic and Pacific coasts are vulnerable, with the exception of the southeastern Atlantic coast. Considering such measures of vulnerability is important for reasons of social justice and can also focus attention on potentially effective risk management interventions unrelated to hazard and exposure that might reduce such vulnerability.

Figure B.1 situates vulnerability as representing a systems view with multiple actors. Vulnerability represents a systems view because the concept focuses on system interactions rather than agency of any particular elements within the system. Vulnerability includes multiple actors because its literature emphasizes that people and communities are not only victims but also active managers of vulnerability (IPCC, 2012, p. 71). While usefully considered an element of the overall framework, the vulnerability literature has established multiple large and credible datasets, and a replicable methodology can be incorporated into a risk management approach.

Barriers

Many cities are vulnerable to impacts arising from climate change, and some portion of those cities are now initiating steps to address these risks through various planning activities. However, many cities and communities in the United States and elsewhere are not well-equipped to mitigate the risks they face from historic climate conditions (Dickson et al., 2012, Chapter 1). The literature on barriers to adaptation offers descriptions and taxonomies of the factors that give rise to the gaps between what might be done and what has been done. Such understanding can usefully inform actions designed to reduce these gaps.

As one important example, Moser and Ekstrom (2010) and Moser and Boykoff (2013) have proposed a framework that uses an idealized adaptation process of understanding, planning, and managing as the basis for its diagnostic process. This understanding, planning, and managing process is consistent with an iterative process of risk management. Recognizing that the adaptation process does not take place in isolation, the Moser and Ekstrom framework acknowledges three "interconnected structural elements": the actors making the decisions, the context in which the decisions are being made, and the object being acted upon through the adaptation process. For each step of the adaptation process, the structural model establishes the source of the barriers by asking such questions as the following: What causes the impediments? How do the actors, context, and the system of concern contribute to barriers (Moser and Ekstrom, 2010, p. 22027)?

---

[2]   Using principal component analysis, HRVI researchers found that 7 of the 30 factors explain 71 percent of the total variability in the data. These are race and class, wealth, elderly residents, Hispanic ethnicity, special needs individuals, Native American ethnicity, and service industry employment.

The answers to these diagnostic questions can then be used to complete a two-item-by-two-item matrix that decision actors may use to identify potential points of intervention, as shown in Figure B.3. The matrix sorts barriers by their temporal origin, defining barriers as contemporary or legacy in nature, and their spatial/jurisdictional origin, defining barriers as being remote or proximate. While remote, legacy barriers may be difficult for current actors in the adaptation process to resolve; a proximate, contemporary barrier is one actors could likely remedy now.

While the Moser and Ekstrom framework evaluates barriers that arise during the adaptation process, Ford and King (2013) take a different approach, proposing a framework to evaluate adaptation readiness, which can be used to identify barriers before the adaptation process begins. The framework's foundation consists of six key factors, derived from the literature on adaptation planning and implementation, which influence decisionmakers' ability to adapt to climate change. (While these were the most common factors that the authors identified, they acknowledged that the list may not be exhaustive for all adaptation scenarios.) These factors are political leadership; institutional organization; availability of usable science to inform decisionmaking; public support; funding for adaptation planning, implementation, and evaluation; and adaptation decisionmaking and stakeholder engagement. To conduct an analysis using this framework, one must first develop indicator variables for each of the six key factors and sources of information that could serve as data to assess how well a goal has been met. Once the data have been collected, scores are assigned to each of the key readiness factors, revealing the varying levels of readiness across the key factors. This framework serves several purposes, as it can be applied to identify areas that could be improved to

Figure B.3
Moser and Ekstrom Barriers to Adaptation

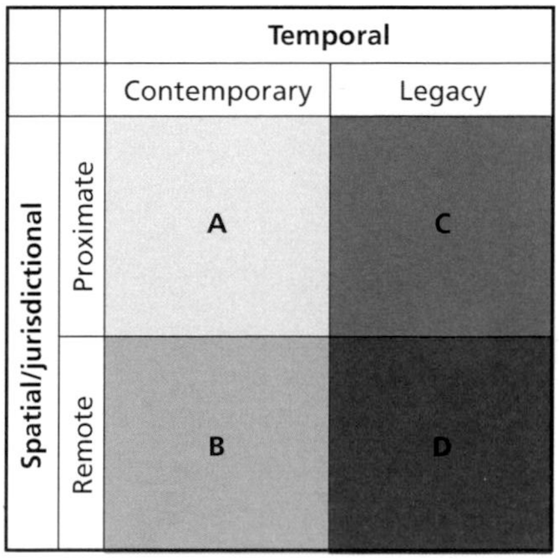

SOURCE: Moser and Ekstrom, 2010.
RAND *RR1144-B.3*

enhance adaptation readiness for a given decisionmaking entity and could also be used to compare adaptation readiness across communities or over time.

The most recent World Bank Development Report (2013) focuses on managing risk. As part of a discussion of implementing effective risk management policies, the report describes a series of barriers (which the report calls "screens") that policymakers might consider in designing an appropriate set of interventions, as shown in Figure B.4. First, incentives for decisionmakers in the private and public sector may cause them to take too much or too little risk. In an urban context, the actions of one agency, such as a flood control department, may cause risks in the jurisdiction of another agency, such as the water quality department, without any formal mechanism to make that risk transfer explicit in the first agency's decisionmaking. Once such disjointed incentives are identified, policymakers may be able to address them. To act properly, even with appropriate incentives, decisionmakers need access to appropriate information. Policymakers may be able to address such information gaps. In some cases, behavioral biases at the organizational or individual level may distort the response to risk. Such policies as education campaigns or introducing new behavioral norms might soften such biases. Finally, decisionmakers may lack adequate resources to manage risks, so that financing mechanisms for long-term investment may become crucial policy responses.

Figure B.1 situates barriers as representing multiple actors lying between a systems-centric and decision-centric view. These barrier frameworks all simultaneously acknowledge that there are actors whose beliefs and actions directly impact climate change adaptation, but that the actions those actors take are either limited or supported by the context in which the decision is made. As expressed by Moser and Ekstrom (2010, p. 22027), their framework aims to be "actor-centric but context-aware."

Resilience

Resilience plays a large and growing role in the literature and practice of responding to climate change and in shaping urban policy and investment more generally. There exist many definitions of the concept, but, fundamentally, resilience is "the capacity of any entity—an individual, a community, an organization, or a natural system—to prepare for disruptions, to recover from shocks and stresses, and to adapt and grow from a disruptive experience" (Rodin, 2014, p. 3).

The modern concept of resilience builds on insights from three fields: engineering, ecology, and systems thinking. In engineering, structures must be designed to weather disruptive events, such as earthquakes or storms, many of which cannot be predicted in any detail. Engineers build resilience by combining strength, flexibility, and redundancy. Resilient structures are strong enough to withstand stresses on them but are also able to bend and shift, while returning to their original form once the stresses subside. In addition, such engineered systems as aircraft, transportation net-

Figure B.4
Barriers to Risk Management from 2014 World Development Report

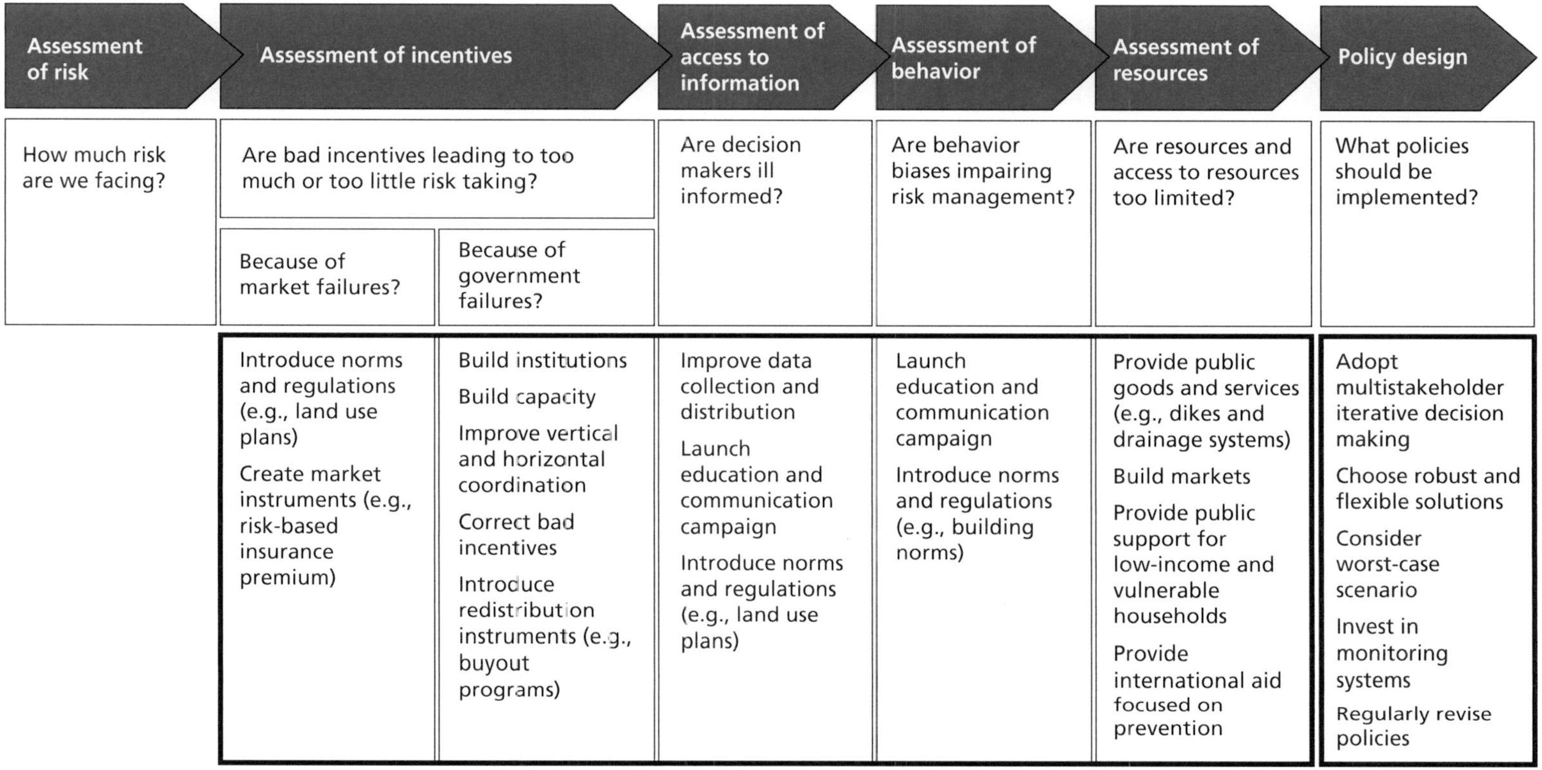

Assessment of risk	Assessment of incentives		Assessment of access to information	Assessment of behavior	Assessment of resources	Policy design
How much risk are we facing?	Are bad incentives leading to too much or too little risk taking?		Are decision makers ill informed?	Are behavior biases impairing risk management?	Are resources and access to resources too limited?	What policies should be implemented?
	Because of market failures?	Because of government failures?				
	Introduce norms and regulations (e.g., land use plans) Create market instruments (e.g., risk-based insurance premium)	Build institutions Build capacity Improve vertical and horizontal coordination Correct bad incentives Introduce redistribution instruments (e.g., buyout programs)	Improve data collection and distribution Launch education and communication campaign Introduce norms and regulations (e.g., land use plans)	Launch education and communication campaign Introduce norms and regulations (e.g., building norms)	Provide public goods and services (e.g., dikes and drainage systems) Build markets Provide public support for low-income and vulnerable households Provide international aid focused on prevention	Adopt multistakeholder iterative decision making Choose robust and flexible solutions Consider worst-case scenario Invest in monitoring systems Regularly revise policies

SOURCE: World Bank, 2013; CC BY 3.0.

RAND *RR1144-B.4*

works, the Internet, and electric grids are all built with redundant components, so that if one fails, others can be used to retain much if not all of the original system function.

Resilience also emerges from the field of ecology. Ecosystems are always changing. For some, the normal range of variation is sufficiently small to make stability a useful concept. A stable ecosystem, similar to an engineered system, has the ability to return to some equilibrium point after a disruption. But many ecosystems exhibit a much larger range of variation. For instance, the mix of vegetation might change dramatically with the seasons. The populations of predators and prey might exhibit large swings. Ecologists use the term *resilience* to mean the property of an ecosystem that, while normally undergoing large changes, is nonetheless able to absorb large shocks while maintaining its intrinsic character. In particular, a resilient system is one that can absorb such shocks without collapse.

Resilience also draws on ideas from systems thinking. Systems dynamics focuses on the behavior of complex systems over time. In particular, such systems thinking highlights feedback loops (both positive and negative) and time delays, which can create chains of cause and effect that differ significantly from what people might naturally infer. As one salient example, people do not often understand the extent to which climate change is a stock problem, rather than a flow problem (Cronin et al., 2009; Sterman, 2011). That is, GHG emissions accumulate in the atmosphere, and the impacts arise after a long delay from the sum total of emissions reductions. This dynamic has profound implications for the viability of a wait-and-see strategy that systems thinking can help illuminate.

As one important attribute, resilience helps integrate consideration of disasters and shocks into a broader theory of system function and change. This connection can prove important because extreme events will be one of the primary ways in which the effects of climate change are felt. In addition, such extreme events may help catalyze desired changes in an urban system. Drawing on its ecological roots, resilience notes that some systems may in fact thrive on shocks. For instance, many forests in the western United States require periodic fires to clear out undergrowth and allow new trees to grow. Without such fires, the system loses resilience because trees become too old and the buildup of fuel creates the potential for catastrophic firestorms. Ecologists capture this idea through the concept of the adaptive cycle, which has four phases. The system begins with rapid growth and then reaches a period of stasis. A disruption then releases the system so that it can reorganize and begin another period of change.

Resilience scholars have applied these ideas to urban settings. For instance, da Silva and Morera (2014, p. 3) define city resilience as: "the capacity of cities to function, so that the people living and working in cities—particularly the poor and vulnerable—survive and thrive no matter what stresses or shocks they encounter." Using a literature review and fieldwork from six cities around the globe, da Silva and Morera identify factors that contribute to city resilience. They offer a three-layer city resilience framework that aims to capture the interdependencies of city infrastructure and power dynamics of city governance across four broad categories of resilience, 12 performance indicators of city resilience, and seven qualities of resilient systems. Da Silva and Morera's seven qualities of resilient systems —reflective, robust, redundant, flexible, resourceful, inclusive, and integrated—are similar to resilience and adaptation requirements found elsewhere in the literature (Nelson et al., 2007; Engle et al., 2013; Moench, 2014). This multilayered system can be used to develop indicators systems that decisionmakers can use to monitor and enhance a city's resilience.

Figure B.1 situates resilience as representing a systems view with multiple actors, consistent with the concept's roots in ecology and systems thinking. Unlike vulnerability and barriers, which are components of risk governance, resilience provides a different framing. While risk management envisions a process of decisionmaking rooted in a unifying measure of risk, resilience summarizes a rich, multifaceted set of desirable qualities. The differences soften somewhat when taking a broader view of risk governance, since an expansive view of the context for managing risks is similar to the broader context required for increasing resilience. Nonetheless, the two have differing foundations and draw from different literatures. Note that it can often prove useful to have a single, easy-to-articulate, all-encompassing goal when seeking to build a common vision and to align the activities of many individuals. As one important advantage, resilience provides such a goal, while at the same time gathering a broad range of attributes under its banner. In this sense, risk management offers a less-focused objective. Overall, it is not yet clear from the literature whether a city would follow fundamentally different paths if it successfully sought resilience than if it successfully practiced a broader view of risk management.

Within a risk governance framing, resilience can be viewed as an important objective of those seeking to manage the risks. This framing of resilience as a goal of risk governance helps address some challenges to the former concept in the literature, in particular those focusing on its normative nature and the related challenge of differentiating good from bad change. A common critique of resilience is that the notion of "bouncing back" fits uneasily with the goal of transformational change. For instance, imagine that a large storm hits a city and washes away vulnerable beachside residential neighborhoods where some families have lived for generations. Does the city better retain its intrinsic character by rebuilding these traditional neighborhoods or by replacing them with new marshlands that provide an innovative mix of recreational opportunities and storm surge defense? In a resilience framework, resolving this choice may present a puzzle. As part of a risk governance process, the particular definition of resilience most suitable for this city, stakeholders, and the public at large becomes a subject of deliberation among the broad group of decisionmakers managing the city's risks.

Precautionary Principle

Quantitative goals often play an important role in climate change policies. For instance, the international community has set a 2°C increase in global mean surface temperatures above pre-industrial levels as the threshold level for dangerous human interference in the Earth's climate (UNFCCC, Copenhagen Accord, 2009, Section 1). Many cities have set quantitative emission reduction targets as a central piece of their GHG reduction programs.

Such quantitative goals play an important role for at least two reasons. First, simple targets can have significant value for motivating and judging performance. As an example, the 2°C goal is easy to communicate and helps focus global attention a problem whose solution requires a vast smorgasbord of policies to address. Without a clear target to motivate action and judge performance, little might get done. Second, quantitative goals can help implement the precautionary principle, which suggests that an activity be avoided unless clear evidence exists that it will not prove harmful. Given the deep uncertainty regarding the impacts of climate change and the potential, however small, for abrupt and cataclysmic change, many decisionmakers favor setting some threshold level of human interference and working backward from those thresholds to set near-term policies.

Such decisionmaking frameworks as guardrails, tolerable windows (Halsnæs et al., 2007), and safe landings (Swart et al., 1998) have been developed to identify GHG mitigation policies consistent with this precautionary approach. For example, the guardrail approach provides an inverse analysis that first defines targets for climate change or climate impacts to be avoided and then determines the range of emissions that are compatible with these targets (Schneider, 2007, p. 802). While these approaches are generally applied to global emissions, they offer tools and frameworks potentially relevant for urban areas undertaking adaptive measures as well.

Figure B.1 situates such precautionary frameworks as representing a decision-actor view with a single actor responsible for policies that achieve the goal. While sometimes offered as an alternative to risk management, precautionary goals and policies to implement them could be regarded as one potential output of a risk management process.

Control Theory and Game Theory

While control theory and game theory are not common frameworks for urban climate programs, they both provide useful perspectives. Control theory is a branch of engineering that focuses on regulating the behavior of dynamic systems with appropriately designed monitoring and feedback mechanisms. An automobile's cruise control system provides a classic example. The car's controller monitors the vehicle's instantaneous speed and adjusts the throttle to maintain the desired rate of travel. While in some ways analogous to the feedbacks that govern the ecological systems envisioned in resilience, control theory focuses on highly engineered constructs whose designers choose the signals to monitor and the responses to those observations.

Figure B.1 thus situates control theory as a single actor lying between a system and decision-actor view. While some literature has applied the framework to global GHG mitigation (Funke et al., 2011), control theory offers two particularly relevant contributions to urban climate risk governance. First, the advent of urban big data that provide real-time monitoring of such parameters as traffic, energy use, water, transit ridership, and air quality can open opportunities for enhanced management of vital, climate-related city functions. Control theory can inform the design of such short time scale (seconds to hours) decision support systems. Second, control theory contains a body of theorems, methods, and experience that can enhance insights and understanding relevant to the design of adaptive decision strategies. These are strategies that evolve over time in response to new information (Lempert et al., 2011). Such adaptive strategies will likely prove central to urban climate risk governance.

Game theory provides a conceptual and mathematical understanding of interacting systems of multidecisionmakers, each pursuing their own interests generally (though not always) with the benefit of expectations about how their own choices, along with other incentives and disincentives, will affect the behavior of other actors. For instance, economists may use game theory to predict how private firms might respond to regulatory requirements imposed by the government in order to evaluate the extent to which proposed regulations might or might not achieve their desired goals. As one example, game theory has seen increased application in the climate literature to suggest how to assemble enduring coalitions of countries committed to global GHG emissions reductions (Bosetti et al., 2013; DeCanio et al., 2013).

Figure B.1 situates game theory as representing a systems view with multiple actors. It is particularly important to urban climate risk governance in understanding how the incentives, information, barriers, and resources summarized in Figure B.4 might affect the choices made by diverse sets of actors in urban settings, how these choices increase or reduce risk, and how alternative interventions might change actors' choices and the resulting risks.

Overview of Existing Climate Adaptation Indicator Systems

We examine indicators that have been developed for measuring adaptation in cities or that have the potential to inform such measurement. We do so, however, with an eye on general applicability to the full range of urban responses to climate change, including mitigation. Despite the limited literature and its potential shortcomings, there are several frameworks and indicators that may hold value for our purposes. Note that this appendix describes these indicators using the language that they or third-party observers use to describe them and makes no attempt to ensure that that language is consistent with the definitions applied in this report.

We discuss these in three groups: national, regional, and urban adaptation indicators; planning indicators; and project and portfolio indicators, summarized in Table C.1. We review national, regional, and local indicators together because they all seek to measure adaptation processes and outcomes across sectors and populations. It seems plausible that regional and national adaptation indicators could be scaled down

Table C.1
Summary of Systems of Indicators Reviewed

Indicators to Assess National, Regional, and Urban Adaptation	Indicators to Assess Adaptation Planning Processes	Evaluation of Adaptation Projects and Portfolios
United Kingdom: National Indicator 188	United Kingdom Evaluation Framework for Public Agency Climate Action Plans	Lamhauge et al., 2012: Review of International Development Agency Practices
Australian Government: Initiative for Local Adaptation Planning	Füssel, 2007: Framework for Assessing Adaptation to Health Risks of Climate Change	Bours et al., 2013: Review of Development Agencies and NGO Practices
United Kingdom's Department for International Development: TAMD	Preston et al., 2011: Evaluation Framework for Adaptation Planning	
United Nations: Evaluating Adaptation Using National Communications in UNFCC	Planned Learning: Based on Swanson et al., 2007, 2010	
Rockefeller Foundation's Resilient Cities Program		

to the city level. We also include indicators of urban climate resilience in this discussion, given the growing emphasis on resilience as a goal for climate change adaptation. There is much literature on resilience in general, not only to climate change but also to natural disasters, economic shocks, terrorism attacks, and other kinds of short-duration disruptions. We focus on indicators that are driven by, or prominently focus on, climate resilience.

We next review indicators that assess planning processes—i.e., the steps taken by public bodies to develop adaptation plans and investments—as distinct from the actions themselves. Adaptation planning can occur at any level: within a single agency, within a sector, city-wide, regionally, or nationally. As such, these indicators hold promise for evaluating cities' adaptation planning processes in different sectors or agencies and in the city as a whole. Planning processes for mitigation include many similar elements.

Finally, we summarize the literature on evaluating adaptation projects and portfolios of projects. This body of work is large, and we have relied on review articles to ensure comprehensiveness. These indicators are often tailored to specific sectors and regions, which may make them difficult to aggregate into a broader set of city-wide indicators.

National, Regional, and Local Adaptation Indicators

Some indicators assess how much progress is being made toward climate change adaptation at the national, regional, and city levels. These indicators assess adaptation in the aggregate—across sectors, population demographics, types of policies and interventions, etc. We have found only a few sets of indicators for assessing adaptation in aggregate:

- National Indicator 188 was the climate change indicator in the United Kingdom's now-defunct Local Government Performance Indicators, by which local governments reported to the national government on progress on key issues.
- Baker et al. (2012) developed a framework for evaluating adaptation plans in response to the Australian Government's initiative in 2008 for local adaptation planning.
- The United Kingdom's Department for International Development's (DfID's) TAMD framework seeks to assess progress toward adaptation in developing countries.
- Gagnon-Lebrun and Agrawala offer a general framework aimed at measuring national progress by evaluating countries' National Communications under the UNFCC.
- The Rockefeller Resilient Cities framework is an effort to understand what makes cities resilient and how resilience can be measured. While the framework is gen-

eral, it is motivated in large part by Rockefeller's efforts to promote resilience to climate change in particular.

United Kingdom's Local Government Performance Indicators National Indicator 188

From 2008 to 2011, the United Kingdom implemented the Local Government Performance Indicators, a set of 198 indicators against which local governments could benchmark their performance in areas of national priority. Among these priorities was adaptation to climate change. National Indicator 188 (NI188) sought to "embed the management of climate risks and opportunities across the local authority and partners services, plans and estates and to take appropriate adaptive actions where required" (Local and Regional Partnership Board, 2010, p. 5). Unlike other indicators in the set that are outcome-based, NI188 was process-based:

> NI188 both recognises that our understanding of the adaptation agenda is not yet sufficient to specify outcomes, but also that climate impacts are local and it is impossible to have a generic outcome indicator at the moment which is applicable to all areas (Local and Regional Partnership Board, 2010, p. 5).

As such, it "measures progress on assessing and managing climate risks and opportunities, and incorporating appropriate action into local authority and partners' strategic planning" (Local and Regional Partnership Board, 2010, p. 4). NI188 required local governments to report on their performance against five process levels, numbered 0 to 4, as shown in Table C.2. For each level, guidance documents described its technical definition, its rationale, aims, a commentary and discussion on aims, and links to other resources.

Unfortunately, we cannot draw conclusions about the effectiveness of these indicators or how local governments used them. In October 2010, the Department for Communities and Local Government instituted sweeping policy changes that removed the National Indicators, arguing that the performance indicators were an inappropriate

Table C.2
Levels of Achievement in National Indicator 188

Level	Definition
0	Getting started
1	Public commitment and impacts assessment
2	Comprehensive risk assessment
3	Comprehensive action plan
4	Implementation, monitoring, and continuous review

SOURCE: Local and Regional Partnership Board, 2010, p. 4; Open Parliament License v3.0.

burden on local governments and prevented them from addressing concerns driven by their constituents (UK Department for Communities and Local Government, 2014). Data-gathering from local governments purportedly continued until March 2011, but the results are not readily available.

Evaluating Adaptation Plans in Southeast Queensland, Australia

In 2008, the Australian government provided grants to city councils to develop climate change adaptation plans. Baker et al. (2012) developed a framework for evaluating adaptation plans and applied it to the plans of seven municipalities in Southeast Queensland, the fastest growing region in Australia and one of the most vulnerable to climate change.

Unlike most evaluation frameworks, this one includes both process and outcome indicators. There are two evaluation categories. First, Baker et al. measure overall progress toward awareness, analysis, and action (AAA). That is, they assess whether cities are (1) aware of the key drivers and consequences of climate change, (2) have begun analysis to integrate climate information and assess local impacts, and (3) have taken actions, through policies and investments, to address risks. They evaluate each AAA category in eight outcome areas shown in Table C.3 in the five categories of water resources; environment, biodiversity, and conservation; urban planning; coastal management; and fire management. They use a 0–4 scoring system to code the data and evidence in the plans, shown in Table C.4. Thus, a plan can receive a maximum of 32 points in each of the three AAA areas, for a total maximum score of 96.

Table C.3
Outcome Criteria for Evaluating Adaptation Plans in Southeast Queensland

Water Resource Planning	Environmental Planning and Biodiversity Conservation	Urban Planning	Coastal Management	Fire Management
C1. Water quantity is maintained or improved	C4. Landscape structure, composition, and function are maintained	C6. Urban heat island effects are minimized or avoided	C7. Impacts of sea level rise and storm surge are minimized	C8. Wildfire risks are managed and impacts are minimized or avoided
C2. Water quality is maintained or improved	C5. Ecosystem, species, and genetic diversity is maintained			
C3. Impacts of flooding are minimized or avoided				

SOURCE: Adapted from Baker et al., 2012, p. 130.

Second, Baker et al., 2012, evaluate the quality of the plans by assessing the eight outcome categories against five quality categories, shown in Table C.5, using the same 0–4 scoring system. Thus, each plan can receive a quality score of at most 20 points in each of the eight outcome areas, for a total score of 160. Their study reports average performance across all seven plans, finding that plans score high on awareness but lower on analysis and still lower on actions. Plan quality in each of the eight outcome areas is also low, receiving less than 50 percent of possible points in every area.

Table C.4
Scoring Criteria for Each Outcome

Code Number	Description
0	No evidence of the criterion throughout the plan.
1	The criterion is acknowledged but lacks further definition and does not provide detail.
2	The criterion is mentioned and includes a moderate level of detail. However, it is entirely descriptive and lacks local application and analysis.
3	The criterion is mentioned and includes a limited level of locally specific application using local climate scenario modeling, exposure, vulnerability and/or risk assessments, maps, local historical data, or fieldwork. However, it is largely descriptive.
4	A detailed analysis of the criterion is provided, and it is addressed in a locally specific manner using a variety of tools, such as vulnerability, exposure and/or risk assessments, maps, fieldwork, GIS analysis and modeling, and local climate scenario modeling.

SOURCE: Baker et al. 2012, p. 131. Used with permission.

Table C.5
Evaluation Criteria for Each Outcome

Plan Component	Concepts
1. Information base	Analysis of current and future conditions in regard to the consequences of climate change. Information included data and analysis of local assets and natural resources, identification of non-climate determinants of vulnerability (e.g., population growth), and vulnerability and risk assessments.
2. Vision, goals, and objectives	Long-term vision of how the community will adapt to climate impacts, including the statement of quantifiable objectives and resource targets in regard to conserving resources under altered climates.
3. Options and priorities	Development, consideration, assessment, and prioritization of alternative climate adaptation solutions
4. Actions	Principles to guide land use decisions to achieve goals, including spacial designs, policies, and/or strategies for implementation
5. Implementation and monitoring	The direction of resources to achieve successful plan implementation and monitoring commitments.

SOURCE: Baker et al., 2012, p. 131. Used with permission.

The authors had the opportunity to informally test the Baker et al. framework by applying it to a local jurisdiction in Los Angeles County. After reviewing the jurisdiction's adaptation planning documents, we met with two officials and interviewed them for about an hour. We then used this information to evaluate their adaption plan using the Baker et al. framework. We found that the jurisdiction, which has completed its assessment phase and is in the process of completing the planning phase of its climate adaptation program, scored highly on each of the five categories in Table C.5. We found the Baker et al. framework generally helpful, but with two important shortcomings. First, the framework offers no guidance on how to determine which outcome criteria ought to be evaluated. Second, the framework does not measure an organization's capacity to implement its adaptation plan. This jurisdiction happened to have significant capacity to implement, due in part to a fortuitous alignment of its existing organizational procedures and authorities with the actions likely required to respond to climate change. This was one of the most promising features of its adaptation efforts. The system of indicators proposed in Chapter Three aims to address these shortcomings.

Tracking Adaptation and Measuring Development Framework

The United Kingdom's Department for International Development (DfID) is the government agency responsible for the country's international aid programs. "Climate and Environment" is one of DfID's six priority areas of work; the first area of focus under this topic is to help the poor adapt to climate change (DfID, 2014). To aid this work, DfID has funded the International Institute for Environment Development (IIED, 2013) to create a framework for measuring climate change adaptation and development in developing countries. The aim of developing the framework was "to see if there was a small number of indicators that could be aggregated across a wide range of adaptation interventions, and that could inform high-level decision making about the use of adaptation resources" (IIED, 2013, p. 4). The resulting Tracking Adaptation and Measuring Development (TAMD) Framework is both process- and outcome-based:

> TAMD is a "twin-track" framework that evaluates the extent and quality of climate risk management (CRM) processes and actions on the one hand (Track 1), and the associated development and adaptation outcomes (and their longer term impacts) "on the ground" on the other (Track 2) (Brooks et al., 2013, p. 8).

Track 1 measures the progress in climate change adaptation processes through changes in institutions, policies, and capacities, while Track 2 measures the outcomes of these processes in terms of changes in well-being, vulnerability, resilience, and security. The TAMD envisions that the Track 1 efforts will be driven by national policy down to local policy, while the Track 2 efforts will be driven by local changes and aggregated up to national impacts. This is shown in Figure C.1.

While the TAMD is flexible and encourages development of locally relevant indicators, it also suggests categories of indicators in each track. The Track 1 indicator categories are shown in Table C.6. Both Track 1 and Track 2 indicators can be scaled to national, regional, and local levels, as shown in Table C.7. TAMD also includes scorecards by which each indicator category can be measured (not shown).

Figure C.1
Schematic of the TAMD's Two-Track Approach to Measuring Adaptation

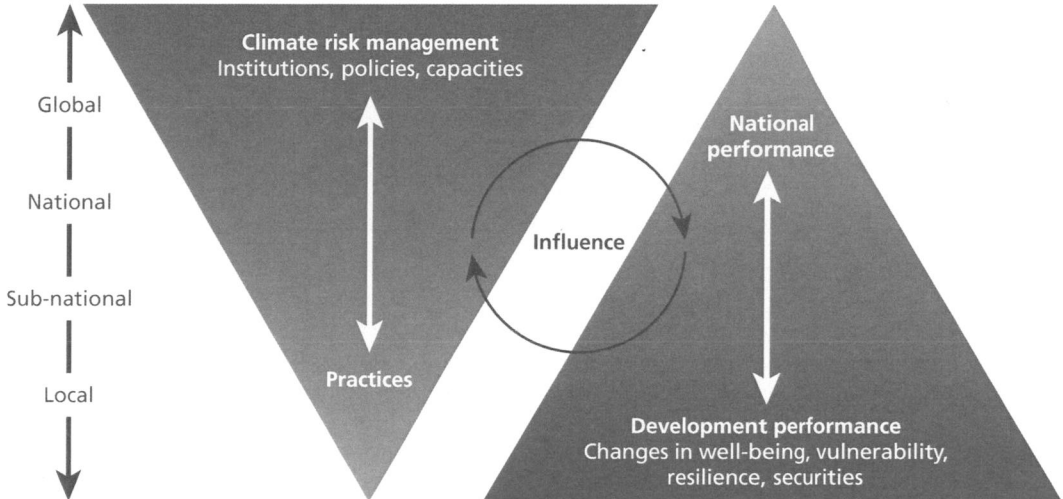

SOURCE: Brooks et al., 2013, p. 10. Used with permission.
RAND *RR1144-C.1*

Table C.6
TAMD Track 1 Indicator Categories

Level	Definition
Indicator 1	Climate change integration into planning
Indicator 2	Institutional coordination for integration
Indicator 3	Budgeting and finance
Indicator 4	Institutional knowledge and capacity
Indicator 5	Climate information
Indicator 6	Uncertainty
Indicator 7	Participation
Indicator 8	Awareness among stakeholders
Indicator 9	Vulnerability/resilience

SOURCE: Brooks et al., 2013. Used with permission.

DfID has conducted an initial assessment of the applicability of TAMD in five countries—Ghana, Kenya, Mozambique, Nepal, and Pakistan—and concluded that TAMD can be applied readily in all of them (IIED, 2013). In the next phase of this work, DfID and IIED will apply the framework in each of these countries (IIED, 2014).

Table C.7
TAMD Indicator Categories

Level	Track 1: CRM	Track 2: Development and Adaptation Outcomes
Global	• Aggregation of national-level performance indicators across countries	• Aggregation of national-level performance indicators across countries
National	• Integration (of CC into planning)—Indicator 1 • Institutional coordination—Indicator 2 • Budgeting and finance (for integration and adaptation)—Indicator 3 • Institutional knowledge (of CC, adaptation, and integration)—Indicator 4 • Use of climate information (to inform planning)—Indicator 5 • Planning under uncertainty (using appropriate information and methodologies)—Indicator 6 • Participation (of relevant stakeholders in national planning processes)—Indicator 7 • Awareness among stakeholders (of climate change, risks, and responses)—Indicator 8	• Aggregation of local/regional data on number experiencing changes in vulnerability and development status • Changes in climate-related economic losses and other impacts (e.g., people affected by climate-related disasters) at national level, in conjunction with data on evolving climate hazards (exposure)
Subnational	• As for national level, adapted for regional contexts	• Aggregation of local data on numbers experiencing changes in vulnerability and development status • Changes in climate-related losses and other impacts at regional/municipal level, in conjunction with data on evolving climate hazards (exposure)
Local	• As for national level, adapted to contexts relevant to key local (formal) institutions, if relevant • Uptake of CRM measures such as risk-spreading mechanisms (financial, livelihood, social)—Indicator 9 • Awareness (of climate risks, trends, prospects, response options)—modified version of Indicator 8 • Climate information (availability, access, use of)—modified version of Indicator 5	• Numbers of people becoming more or less vulnerable, as measured by a variety of context-specific vulnerability indicators • Changes in poverty and other standard development indicators, complemented by or "normalized" with respect to changes in climate hazards (exposure)

SOURCE: Brooks et al., 2013, p. 15. Used with permission.
NOTE: CC = climate change.

Evaluating Adaptation in UNFCCC National Communications

Countries that are signatories to the UNFCCC are required to submit national reports on their implementation of the Convention to the Conference of Parties. While the initial National Communications (NCs) focused largely on GHG emissions and broad discussions of risk, they increasingly include countries' efforts on climate change adaptation.

Gagnon-Lebrun and Agrawala (2007) develop and apply a framework for evaluating national-level progress on adaptation, as documented in the NCs from developed countries. As shown in Table C.8, they describe eight steps along which climate change adaptation proceeds, divided into three categories: climate change and impacts assessment, intention to act, and adaptation actions. They evaluate each NC at each step according to the scope of issues covered (extensive, limited, or lacking) and depth of coverage (detailed, generic, limited but with references to comprehensive studies, or lacking). Based on these criteria, they divide countries into three categories of progress: (1) early to advanced stages of impact assessment; (2) advanced impact assessment, but slow development of adaptation responses; and (3) advanced impact assessment and moving toward implementing adaptation. They find that most developed countries fall into category two. They note that, at the time of their study, "No developed country has yet formulated a comprehensive approach to implementing adaptation and the 'mainstreaming' of such measures within sectoral policies and projects, although the UK might be coming close" (Gagnon-Lebrun and Agrawala, 2007, p. 401).

While Gagnon-Lebrun and Agrawala developed their framework for national evaluations and apply them to national communications, the processes and criteria are general enough that they could be readily applied to local, urban adaptation processes as well. For example, the process steps they outline (shown in Table C.8) largely overlap with the process steps defined by the UK National Indicator 188, which was designed for measuring adaptation progress of municipalities.

Table C.8
Climate Change Adaptation Process

| Climate Change and Impacts Assessments | Adaptation | |
	Intention to Act	Adaptation Actions
• Historical climatic trends • Climate change scenarios • Impacts and risks assessments	• Identification of adaptation options • Mention of existing policies synergistic with adaptation	• Establishment of institutional mechanisms • Formulation of policies/modification of existing policies • Explicit incorporation of adaptation in projects

SOURCE: Adapted from Gagnon-Lebrun and Agrawala, 2007, p. 398.

Rockefeller Resilient Cities Framework

In May 2013, the Rockefeller Foundation announced its "100 Resilient Cities Centennial Challenge," in which it would select 100 cities to receive support to address future threats and shocks. Nearly 400 cities applied by the October 2013 deadline, and in December 2013, the foundation announced the first 33 selected cities (Rockefeller Foundation, 2014). Complementing this effort, the Rockefeller Foundation is developing a framework for measuring urban resilience and identifying resilient cities (Arup, 2014, p. 1):

> What and who makes a city resilient—and not just liveable now or sustainable for the long term—has become an increasingly critical question, one we set out to answer in late 2012 with our partners at Arup through the creation of a City Resilience Index (Arup, 2014, p. 1).

While climate change resilience is a key driver behind the development of this framework, it is aimed at measuring resilience to other shocks, including financial shocks, terrorism, pandemics, and slow-moving chronic stresses.

The framework recognizes several general qualities of resilient cities, summarized in Table C.9, which it applies to four areas of urban resilience: health and well-being, economy and society, urban system and services, and leadership and strategy. Three indicators measure resilience in each of these categories for a total of 12 resilience indicators, listed in Table C.10. Each indicator has an additional four to five subindicators.

Table C.9
Qualities of Resilient Cities

Quality	Description
Reflective	Reflective systems are accepting of the inherent and ever-increasing uncertainty and change in today's world.
Robust	Robust systems include well-conceived, constructed and managed physical assets, so that they can withstand the impacts of hazard events without significant damage or loss of function.
Redundant	Redundancy refers to spare capacity purposely created within systems so that they can accommodate disruption, extreme pressures or surges in demand.
Flexible	Flexibility implies that systems can change, evolve and adapt in response to changing circumstances.
Resourceful	Resourcefulness implies that people and institutions are able to rapidly find different ways to achieve their goals or meet their needs during a shock or when under stress.
Inclusive	Inclusion emphasizes the need for broad consultation and engagement of communities, including the most vulnerable groups.
Integrated	Integration and alignment between city systems promotes consistency in decision-making and ensures that all investments are mutually supportive to a common outcome.

SOURCE: Adapted from Arup, 2014, p. 5.

Table C.10
Indicators of Urban Resilience

Category	Indicator
Health and Well-Being	**1. Minimal human vulnerability.** Indicated by the extent to which everyone's basic needs are met.
	2. Diverse livelihoods and employment. Facilitated by access to finance, ability to accrue savings, skills training, business support, and social welfare.
	3. Adequate safeguards to human life and health. Relying on integrated health facilities and services and responsive emergency services.
Economy and Society	**4. Collective identity and mutual support.** Observed as active community engagement, strong social networks, and social integration.
	5. Social stability and security. Including law enforcement, crime prevention, justice, and emergency management.
	6. Availability of financial resources and contingency funds. Observed as sound financial management, diverse revenue streams, the ability to attract business investment, adequate investment, and emergency funds.
Urban System and Services	**7. Reduced physical exposure and vulnerability.** Indicated by environmental stewardship, appropriate infrastructure, effective land use planning, and enforcement of planning regulations.
	8. Continuity of critical services. Indicated by diverse provision and active management, maintenance of ecosystems and infrastructure, and contingency planning.
	9. Reliable communications and mobility. Indicated by diverse and affordable multi-modal transport systems and information and communication technology networks and contingency planning.
Leadership and Strategy	**10. Effective leadership and management.** Involving government, business, and civil society and indicated by trusted individuals, multistakeholder consultation, and evidence-based decisionmaking.
	11. Empowered stakeholders. Indicated by education for all and access to up-to-date information and knowledge to enable people and organizations to take appropriate action.
	12. Integrated development planning. Indicated by the presence of a city vision, an integrated development strategy, and plans that are regularly reviewed and updated by cross- departmental working groups.

SOURCE: Adapted from Arup, 2014, p. 5.

For example, the subindicators for diverse livelihoods and employment are livelihood opportunities, skills and training, development and innovation, and access to financial assistance. (We have not listed all subindicators; Arup, 2014, pp. 8–13).

The framework is not yet complete and has not yet been applied in practice. As a next step, the foundation intends to develop a set of 128 additional variables that will help measure resilience along these indicators and subindicators (Arup, 2014, p. 1).

Summary of National, Regional, and Local Adaptation Indicators

Table C.11 summarizes several features of the frameworks for national, regional, and local metrics that we reviewed. For each, the table notes the type of institution(s) that developed it, the level of governance for which it is intended, whether it is process-based or outcome-based, and the extent to which it has been applied in practice.

First, we observe that different types of institutions have developed these different indicators, reflecting that evaluating climate change adaptation is of concern to a variety of actors: governments, NGOs, stakeholders, and the academic and research community. Second, while the geopolitical scale varies from local to national, nearly all indicators frameworks have a local component or are general enough to be applied to local adaptation efforts. Thus, these frameworks may each hold promise for measuring cities' progress in climate change adaptation. Third, all of the climate change adaptation indicators include process-based indicators, while some additionally include outcome-based indicators. This is consistent with our discussion of the many challenges to developing outcome-based indicators. The Rockefeller Resilient Cities is by definition outcome-based because it measures urban resilience—an outcome of climate change adaptation.

None of the frameworks has been widely used in practice for a number of different reasons. The UK National Indicator has been discontinued for political reasons, while the frameworks developed in the academic community seem to have been applied in one-off studies. It is not clear whether they will continue to be applied. The TAMD framework seeks to measure progress at all scales (locally to nationally),

Table C.11
Summary of National, Regional, and Local Metric Frameworks

Metric	Developer	Geopolitical Scale	Process-Based or Outcome-Based	Application/Use
UK National Indicator 188 (Board, Local and Regional Partnership, 2010)	National government	Local	Process	Unknown, discontinued
Baker et al., 2012	Research/academic	Local	Process and outcome	Seven Australian municipalities
TAMD, 2014	National government and NGO	Local to national	Process and outcome	Initial assessment in five developing countries
Gagnon-Lebrun and Agrawala, 2007	Research/academic	National	Process	39 developed countries' UN National Communications
Rockefeller Resilient Cities (Rockefeller Foundation, 2014)	NGO	Local	Outcome	N/A; under development

involves both process and outcome indicators, and has articulated criteria to assess progress along each metric. DfID and IIED are in the process of piloting TAMD in five developing countries. The Rockefeller Resilient Cities framework is still under development.

Adaptation Planning Indicators

Another class of indicators was developed expressly to assess adaptation plans and planning processes that subsequently lead to climate change adaptation policies, investments, and other actions. These do not measure the degree of adaptation that has occurred; rather, they assess the quality of the planning processes that may lead to action.

We reviewed the following evaluation frameworks for planning:[1]

- The United Kingdom has developed an evaluation framework for assessing climate change adaptation reports submitted by public agencies under the Climate Change Act of 2008. The framework evaluates qualities of the agencies' risk assessments.
- Füssel offers a framework for assessing adaptation to health risks of climate change. We include it in our review because most of the indicators are relevant to adaptation planning in general and not limited to health outcomes.
- Preston developed a systematic evaluation framework for adaptation planning and has applied it to 57 adaptation plans from Australia, the United Kingdom, and the United States.

Evaluation of Risk Assessments under the Climate Change Act 2008

The UK's Climate Change Act 2008 gives the UK Department for Environment, Food and Rural Affairs (Defra) authority to require public service organizations, such as the Environment Agency, water companies, and electricity distributors, to produce climate change adaptation reports. The reports are used to assess how climate change impacts the organization's functions and how the organization proposes to adapt to climate change. Cranfield University developed a framework that Defra could use to evaluate the risk assessment component of the adaptation reports. The evaluation framework consists of eight attributes related to the use of data, managing uncertainties, adaptation priorities and outcomes of the assessment, and monitoring of effectiveness, as shown in Table C.12. Each of the attributes includes numerous subattributes that are used to score the attribute as being not present, partially complete, complete, or com-

---

[1] Note that the "plan quality" portion of the framework developed by Baker et al. (2012) for assessing climate change adaptation plans could also be discussed as a planning metric. For brevity, we do not repeat that discussion here.

Table C.12
Evaluation Framework for Risk Assessments Under the UK Climate Change Act 2008

Number	Attribute
1	Climate change risk assessment is a clear component of corporate risk appraisal.
2	Climate change risk assessment enables the Reporting Authority to make evidence based decisions on adapting to climate change.
3	Demonstrable use of relevant and appropriate data, information, knowledge, tools and methodologies
4	Climate change risk assessment and adaptation measures explicitly consider uncertainties.
5	Climate change risk assessment generates priorities for action.
6	Climate change risk assessment identifies opportunities (where applicable).
7	Clear demonstration of flexible adaptation measures
8	Monitoring and evaluation of adaptation effectiveness

SOURCE: Drew et al., 2010; Crown Copyright, Open Government License v3.0.

plete and fully integrated (Drew et al., 2010). Table C.13 shows the subattributes for Attribute 4, "Climate change risk assessment and adaptation measures explicitly consider uncertainties."

Defra received nearly 100 reports in the first round of reporting, from December 2010 to December 2011.[2] Reviewers of these reports note that "the Evaluation Framework has for the most part proved to be a valuable tool, providing objective, standardised assessments of the reports" (Centre for Environmental Risks and Futures, 2012, p. 12). The next round of reporting will take place from July 2013 to 2016, and a summary of evidence of outcomes will be included in the next Climate Change Risk Assessment report, due for release in January 2017 (UK, 2014).

Evaluating Adaptation to Climate Change Health Risks

A number of evaluation frameworks have been proposed in the academic literature without being associated with specific government policies or programs. Füssel (2008), for example, offers 14 criteria for evaluating adaptation planning processes and applies these criteria to national and international guidance documents (Table C.14). However, the criteria do not include specific indicators, indicators, or thresholds that would help assess whether a particular planning process meets or fails to meet a particular criterion (Füssel, 2008).

Füssel (2008) applies these criteria to several climate change guidance documents and models for environmental epidemiology: IPCC Technical Guidelines, the United States Country Studies Program (USCSP) Handbook, the UNEP Handbook,

---

[2] Reports are archived at Department for Environment, Food and Rural Affairs, 2012.

Table C.13
Subattributes and Evidence for Attribute 4 of the Evaluation Framework for Risk Assessments

Subattribute	Not Complete	Partially Complete	Complete	Complete and Fully Integrated
4.1. Reporting Authority's risk assessment includes a statement of the main uncertainties in the evidence, approach and method used in the adaptation plan and in the operation of the organization.	No evidence identified	Identification of main uncertainties in the evidence, approach and method, but little/ no consideration of how this affects the overall risk assessment	Explicit discussion of the key uncertainties in the evidence, in the risk assessment approach, with implications for the risk assessment findings	Exploration of the sensitivities of the risk assessment to key uncertainties, with alternative actions for priority risks that are vulnerable to underlying uncertainties
4.2. Reporting Authority's adaptation responses explicitly account for uncertainties and interdependencies of actions, including the actions of others on the adaptation plan.	No evidence identified	Some indication of how the adaptation response can deal with uncertainty, and identification of other organizations that may impact on adaptation response	Good coverage of how the adaptation response is robust to uncertainties, and discussion of the extent to which management of the Reporting Authority's risks are contingent on other organizations' actions	Full coverage of how the adaptation response is robust to uncertainties, and exploration of the sensitivities of others' actions on the Reporting Authority's risks, together with plans to address these
4.3. Reporting Authority's adaptation plan includes a clear statement of assumptions which are well evidenced and justified.	No evidence identified	Statement of assumptions within adaptation plan but not how these impact on the resulting actions	Rationale for the assumptions made, set within an organizational context, so establishing the credibility of assumptions, and discussion of how they impact on the findings and how they can be addressed	Exploration of the sensitivity of adaptation plan to underlying assumptions

SOURCE: Drew et al., 2010, p. 21; Crown Copyright, Open Government License v3.0.

the United Kingdom Climate Impacts Programme Framework for Climate Adaptation, the United Nations Development Program Global Environment Fund's Adaptation Policy Framework, and the World Health Organization—Health Canada Assessment Framework. Füssel (2008) finds that none of the guidance documents adequately

Table C.14
Criteria for Assessing Adaptation to Climate Change Health Risks

Number	Criteria
1	Clear procedural structure*
2	Flexible assessment procedure*
3	Prioritization of assessment efforts
4	Identification of key information needs*
5	Inclusion of key stakeholders*
6	Choice of relevant spatial and temporal scales
7	Balanced consideration of current and future risks*
8	Management of uncertainties*
9	Policy guidance in the absence of quantitative risk estimates*
10	Prioritization of adaptation actions*
11	Mainstreaming of climate adaptation*
12	Cross-sectoral integration*
13	Disease-specific methods and tools
14	Assessment of key obstacles to adaptation*

SOURCE: Adapted from Füssel, 2008.

NOTE: These criteria are specifically aimed at human health effects, but most are more broadly applicable (denoted with an asterisk).

address all criteria, suggesting that assessments of plans must use guidance from several documents simultaneously.

Preston et al.'s Framework for Evaluating Adaptation Planning

Consistent with our findings in this review of monitoring frameworks, Preston et al. (2011, p. 411) argue that many evaluation frameworks and criteria are ad hoc and unsystematic:

> The lack of consensus among guidance instruments highlights the fact that a systematic approach to monitoring and evaluation for climate change adaptation has yet to emerge, and the capacity to undertake such monitoring and evaluation and incorporate it into adaptation policy is lacking.

In response to these shortcomings, Preston et al. (2011) offer a new framework, arguably developed with a more rigorous methodology. The framework draws on logical framework analysis (LFA):

The LFA approach incorporates the analysis of (a) the relationships among program goals and objectives; (b) the activities by which those objectives may be realized; (c) the inputs and resources required to undertake those activities; and (d) the outputs that emerge from the execution of identified activities. Each of these may be associated with one or more indicators as well as a host of underpinning assumptions (Preston et al., 2011, p. 414).

Corresponding to these steps, the framework defines four stages of adaptation planning: (1) goal-setting (to capture relationships between goals and objectives), (2) stock-taking (to identify inputs), (3) decisionmaking (to allocate inputs and resources), and (4) implementation and evaluation (to realize the outcomes). Preston et al. then reviewed 20 guidance documents for adaptation planning to identify the 19 process-based evaluation criteria shown in Table C.15. A score of 0, 1, or 2 is possible for each criterion, indicating no evidence, evidence with limitations, and strong evidence, respectively. Thus, the total possible score is 38.

Preston et al. (2011) examined a variety of characteristics across the 57 adaptation plans. They report that the average score was 14 (37 percent of the total), ranging from a low of 6 out of 38 (16 percent) to a high of 22 out of 38 (61 percent). They found that local plans scored slightly higher than regional or national plans, hypothesizing that this may suggest that it is easier to develop detailed plans when the geopolitical scope is limited. However, in the last several years, the quality and depth of adaption plans has improved significantly (see Georgetown Climate Center, 2016, for a full listing).

Preston et al. (2011) also found that many adaptation plans are politically risk-averse. These plans avoid mention of adaptation actions that seek to manage unavoidable loss, such as spreading risk through insurance. The authors hypothesize that the public's expectation of government is to protect and preserve, and that institutions are loath to acknowledge that not all losses can be avoided.

Finally, Preston et al. (2011) found that many adaptation actions cited in adaptation plans were no- or low-regret and included actions that governments would take anyway for other environmental or risk management reasons. They suggest that this near-term view may have long-term consequences (p. 425):

> This emphasis on short-term benefits may hinder the implementation of adaptation actions needed to secure long-term resilience, such as enhanced investments in long-lived infrastructure to ensure it is robust against potential climate change decades in the future (Fankhauser et al., 1999).

Summary of Planning Indicators

In this section, we reviewed three frameworks for indicators for evaluating the quality and characteristics of climate change adaptation plans and the processes used to develop those plans. The first is the United Kingdom's evaluation framework for assessing public agencies' adaptation reports. While Defra has received and evaluated many

Table C.15
Evaluation of Adaptation Planning Stages and Criteria to Evaluate Them

Adaptation Stage	Stage Descriptions	Adaptation Processes	Criteria Descriptions
Goal-Setting (Goals, Objectives, Purpose)	Establishing what decision-makers seek to achieve through adaptation and how performance with respect to obtaining goals will be determined	Articulation of objectives, goals and priorities	Establishing the objectives, goals and priorities for adaptation
		Identification of success criteria	Consideration of what successful adaptation will look like and how it will be measured
Stock-Taking (Inputs)	Assessing institutional assets and liabilities that facilitate or hinder adaptation planning and policy implementation. As such, this stage effectively represents an assessment of adaptive capacity. However, to further discriminate between different components of adaptive capacity, this stage was conceptualized as assessment of five stocks of capital relevant to adaptation, based upon the sustainable livelihoods literature (Ellis, 2000; Nelson et al., 2005, 2007; Iwanski et al., 2009).	Assessment of human capital	Consideration of the existing skills, knowledge and experience of individuals responsible for adaptation planning and implementation
		Assessment of social capital	Consideration of the existing governance, institutional and policy contexts for adaptation, including the capacity and entitlements of those institutions, organizations and businesses responsible for designing, delivering and implementing adaptation measures
		Assessment of natural capital	Consideration of natural resource stocks and environmental services which are sensitive to climate and/or integral in the management of climate risks
		Assessment of physical capital	Consideration of material culture, assets and infrastructure that is sensitive to climate and/or integral in the management of climate risks

Table C.15—continued

Adaptation Stage	Stage Descriptions	Adaptation Processes	Criteria Descriptions
		Assessment of financial capital	Consideration of stocks and flows of financial resources and obligations within and among individuals and institutions including cash revenue, credit and debt and mechanisms for financial risk management
Decisionmaking (Activities)	Processes associated with determining what adaptation policies and measures are appropriate. This stage encompasses a variety of tasks, from engaging with stakeholders about preferred adaptation responses, assessment of climate and nonclimate system drivers, assessment of impacts, vulnerability and risk and the prioritization of different adaptation options and their harmonization with existing policy structures.	Stakeholder engagement	Engagement of relevant stakeholders and communities throughout the adaptation process
		Assessment of climate drivers	Consideration of historical climate trends, current climate variability and future climate projections
		Assessment of nonclimate drivers	Consideration of variability and trends in other environmental and socioeconomic factors relevant to the system of interest
		Assessment of impacts, vulnerability, and/or risk	Assessment of the impact of changes in climate, vulnerability or resilience to those changes and the relative importance of climate and nonclimate risks
		Acknowledgement of assumptions and uncertainties	Transparency about the assumptions made to establish those impacts and risks and the uncertainties involved in their estimation
		Options appraisal	Identification and comparison of different adaptation options and a means for selecting between them

Table C.15—continued

Adaptation Stage	Stage Descriptions	Adaptation Processes	Criteria Descriptions
		Exploitation of synergies	Identification of where opportunities exist to implement adaptation in a manner that promotes synergies with existing policies or plans, including mitigation
		Mainstreaming	Identification of ways in which climate change adaptation can be institutionalized or embedded into existing or new policies and plans
Implementation and Evaluation	Processes associated with the implementation of preferred adaptation options which may include communication, the removal of barriers and the assignation of roles and responsibilities. In addition, this stage also includes downstream processes associated with monitoring and evaluation of implemented actions.	Communication and outreach	Communication and dissemination of adaptation plans and any downstream outcomes to the appropriate stakeholders and communities
		Definition of roles and responsibilities	Establishing who is responsible for different aspects of an adaptation strategy
		Implementation	Establishing the mechanisms that will allow implementation of adaptation measures
		Monitoring, evaluation and review	Establishing a system of monitoring and evaluation that allows the performance of adaptation to be assessed against success criteria and for review of inputs and procedures

SOURCE: Preston et al., 2011, pp. 416–417. Used with permission.

reports using this framework, we have not been able to find the results and therefore cannot draw conclusions about the usefulness of the framework. The second is Füssel's framework for assessing health risk adaptation. While the criteria are largely transferable to nonhealth issues, the framework does not include specific indicators or indicators.

The third, Preston et al.'s framework, is perhaps most promising and relevant for climate change adaptation in cities. It pays particular attention to methodological rigor and is grounded in an LFA approach. The authors also apply it to a large number of adaptation plans from several countries. Like other metric frameworks, the authors describe adaptation planning as a linear four-step process (goal-setting, stock-taking, decisionmaking, and implementation and evaluation) that is consistent with the logical framework analysis, and they offer climate change-specific indicators for evaluating progress within those steps. However, they extensively discuss issues concerning risk governance: the need for integrating climate change adaptation planning into more complex institutional, organizational, and governance contexts and the use of evaluations of the planning process to describe the shortcomings of adaptation plans in this regard. This suggests that the framework may be suitable to integration into a more complex framework for adaptation planning that takes these interactions into account.

Project and Portfolio Indicators

The international development community has spearheaded much of the work on adaptation indicators, largely to monitor and evaluate individual projects and portfolios of projects that they fund (Bours et al., 2013). These indicators focus on specific project impacts and outcomes and may complement broader indicators for adaptation in cities by helping to assess projects and policies that cities undertake as part of their broader adaptation efforts. They may also offer insights on what adaptation looks like in practice, particularly in terms of implementation.

Nearly every major development organization has created its own metric frameworks for monitoring and evaluating climate change adaptation projects. Several researchers have recently reviewed these frameworks for indicators. Therefore, we provide a brief summary of the second-order literature with examples but do not provide an in-depth review of the larger set of such indicators.

Lamhauge et al.'s Review of Indicators from International Development Agencies

Lamhauge et al. (2012) reviewed commonalities in monitoring and evaluation frameworks from six international development agencies. They found that most agencies use Results-Based Management (RBM), which has become mainstream in the development community and implements a logical framework approach for project management. RBM uses program logic models to articulate how project inputs and activities translate into results, defined as outputs, outcomes, and impacts (ADB, 2006). Table C.16 provides an example of a "logframe" (logical framework) from a Swiss Agency for Development and Cooperation project on building community awareness of climate change.[3] Like other frameworks, the use of indicators for measuring outputs, outcomes,

---

[3] A logframe is a matrix that implements the components of a logical framework analysis. It structures the main elements in a logical framework analysis and notes the linkages between them. A standard logframe has as rows the goals, purpose, objectives, outcomes, outputs, and activities for a project or initiative. The columns assess the outputs and key activities, indicators, means of verification, and important assumptions of each row (USAID, 2003, p. 2).

Table C.16
Logframe from the Swiss Agency for Development and Cooperation (SDC)

Level	Description	Indicator	Means of Verification	Assumptions
Output 2.2	Community aware of localized climate change information and have access to advisory services			
Activity 2.2.1	Test and establish agromet stations (incl. soil moisture, hydrological parameters, etc.) and water budgeting tools	1. Optimal number of agromet stations established to service project villages	1. Monitoring reports 2. Agromet data 3. Documented protocols and tools	The required information regarding meteorological data/ weather conditions/ climate change is available and accessible
		2. Protocols and tools for water-budgeting developed		
Activity 2.2.2	Risk reduction strategies and measures for slow and rapid disaster events developed and advisories generated	1. Local disaster management plans exist and put in place	Documented DRR protocols exist	
		2. Disaster management committees at village level are in place	Monitoring reports	
		3. No. of advisories on water use, crop planning and management, pest management, etc., issued	Advisories	Insurance companies are willing to partner WOTR and develop suitable products
		4. No. and type of (disaster risk reduction) instruments—e.g., insurance instruments promoted	Insurance products	
Activity 2.2.3	Integrate indigenous knowledge and scientific knowledge toward climate change preparedness (disaster preparedness, early warning systems, etc.)	Methodology and mechanisms developed for integration of [indigenous knowledge] with scientific knowledge	Relevant documents	Various experts appreciate the need for [indigenous knowledge] integration and agree on methodology
Activity 2.2.4	Continuously monitor emerging data from national and international studies and collaborate with NDMA and others	Various desk studies/ synthesis reports available and number of exchanges/ meetings	Synthesized reports	Meaningful disaggregated data and studies available and accessible

SOURCE: Lamhauge et al., 2012. Used with permission.

NOTES: DRR = disaster risk reduction; NDMA = National Disaster Management Authority; WOTR = Watershed Organisation Trust.

and impacts is central to the approach. The authors assess common indicators used by these agencies in each of five adaptation areas: (1) climate risk reduction; (2) policy and administrative management for climate change; (3) education, training, and awareness on climate change; (4) climate scenarios and impact research; and (5) coordination on climate change measures and activities across relevant actors. Table C.17 shows indicators common to risk reduction projects. The authors offer modest conclusions and lessons learned across these six frameworks, such as the following:

Table C.17
Common Indicators on Risk Reduction

Indicators	CIDA	DFID	DGIS	JICA	SDC	Sida
No. of households/ communities participating in afforestation/improved agricultural practices/ watershed management			X	X		X
Area of afforestation (m²/ ha)				X		X
Impact of flood (no. of people affected, inundation depth, duration, value of flood damage)				X		
No. and type of DRR instruments (e.g., insurance instruments) promoted					X	
Early warning system in place			X			
Construction of climate-proof infrastructure						X
Percentage of population with improved and sustainable access to water sources		X	X			X
No. of (people benefiting from) water, livestock, and natural risk management projects	X	X	X		X	X
No. of households that seek out, test, adapt, and adopt ideas and practices that strengthen their livelihoods	X	X	X			

SOURCE: Lamhauge et al., 2012. Used with permission.

- The six agencies most often use RBM and LFA for their monitoring and evaluation frameworks.
- Qualitative, quantitative, and binary indicators should be used, regardless of the type of adaptation activity being undertaken.
- Defining baseline values and the project scale are important considerations for choosing indicators.

Lamhauge et al. (2012) also raise a (largely unanswered) question of how these individual indicators can and should be used to capture overall adaptation progress:

> [T]he different approaches used by the agencies—particularly in the context of climate risk reduction—raise the question of whether to use detailed indicators corresponding to every component of an intervention or if a more aggregate measure that captures reduction in overall climate vulnerability is preferable. The answer for this is likely to depend on the type and the scale of the activity. For risk reduction measures, a general assessment on vulnerability may be more appropriate than, for example, for training activities aimed at increasing people's adaptive capacity through the introduction of new livelihood activities (Laumhauge et al., 2012, p. 43).

The answer to this question regarding detailed versus aggregate indicators may provide insights on how these project- and portfolio-level indicators can be used to measure adaptation progress in aggregate in cities.

Bours et al.'s Review of Indicators from Development Agencies and Other NGOs

Bours et al. provide an annotated summary of 16 monitoring and evaluation frameworks specifically aimed at international development projects and programs from nearly as many development organizations and NGOs. They found that earlier frameworks were conceptually strong but weak and oversimplified in practical application. They report a trend toward aggregating indicators to report on higher-level outcomes. This may be necessary for transforming project and portfolio indicators into city-wide measures of adaptation. However, they argue that "there is concern regarding considerable methodological pitfalls and challenges when transforming qualitative, local, participatory data into quantitative targets in this way" (Bours et al., 2013, pp. 60–61). They further note that while many evaluation frameworks press for process-based indicators, more recent frameworks focus on outcome indicators rather than process indicators. The authors hypothesize that this reflects the difficulty of aggregating process-based indicators. Finally, the authors question whether the monitoring and evaluation frameworks fail to identify maladaptation—that is, projects may score well with respect to a framework's indicators but ultimately result in incorrect adaptation or do not meaningfully contribute to adaptation. To mitigate the risk of maladaptation, the risk governance framework proposed in Chapter Three emphasizes the need for iterative risk management, learning, and adaptive planning.

Summary of Project and Portfolio Indicators

Most major development and aid organizations have developed a monitoring and evaluation framework to assess the climate change impacts of their projects. These frameworks draw extensively on the RBM and LFA approaches that have become standard for general project evaluation in this community. These frameworks may hold promise for cities as they seek to evaluate specific climate change adaptation projects. These frameworks are not yet designed to provide indications of aggregate progress in climate change adaptation at the local, regional, or national level:

> [There is a] need to situate the evaluation of specific interventions within broader country objectives. In the context of adaptation, this would mean complementing individual project and programme evaluations with overall assessments of trends in countries' vulnerability to climate change. A framework for linking individual assessments with national level assessments could help to broaden the focus from the means of achieving outcomes (individual interventions) to the desired end result (countries' becoming less vulnerable to climate change). By doing so, the combination of country-level monitoring and project level M&E should highlight the issues of whether the overall level of action is sufficient, how the distribution of vulnerability is changing and whether the composition of interventions is coherent. There is a need for further research to operationalise this type of approach (Lamhauge et al., 2012, p. 44).

This may prove to be one of the challenges in cities using these indicators in practice.

Overview of Responses to Climate Change in Urban Areas of the United States

For over a decade, local governments across the United States have been developing climate change mitigation plans to achieve reductions in GHG emissions. Mitigation actions in urban areas typically focus on energy conservation, low-carbon energy production and use, design, and energy use in commercial and residential buildings, transportation, industry, and waste management. Mitigation actions intended to reduce concentrations of GHG in the atmosphere may either reduce sources or enhance sinks.

More recently, growing concerns about the impacts of climate change to people and assets (with or without emissions reductions) have motivated local governments to undertake climate change adaptation planning as well (Aylett, 2014). Adaptation is not a substitute for efforts to reduce GHG emissions, but rather an acknowledgment that climate change is already under way, regardless of the success of mitigation or lack thereof. With the large inertia in the global climate system, the benefits of mitigation are unlikely to be felt for some time. Therefore, at a minimum, cities and regions are compelled to plan for change over the next several decades.

Adaptive actions may include everything from sea walls, barriers, and bulkheads to beach nourishment, elevation of buildings, and rezoning. Many of these actions involve construction, reconstruction, or reconfiguration of infrastructure. They could cost the United States between tens and hundreds of billions of dollars annually (Sussman et al., 2014).[1] Many of these adaptation efforts will occur in urban areas that encompass multiple local jurisdictions, many separate state and local agencies, federal government assets, and sometimes several states. As the scale and urgency of climate change impacts grow, some urban areas are initiating adaptation planning, often under the banner of increasing resilience, with encouragement from the federal government (HUD, 2014b). However, thus far, the extent of comprehensive planning and implementation remains limited in all but a few areas.

---

[1] Sussman et al. (2014) report that estimates vary widely because they use different methodologies and examine costs over different time scales, assumptions about climate change, and aggregation levels (regional, by sector, etc.), but in general, costs are cumulative through the year 2100.

The 2014 National Climate Assessment (Bierbaum et al., 2014, p. 671) concluded: "Substantial adaptation planning is occurring in the public and private sectors and at all levels of government; however, few measures have been implemented and those that have appear to be incremental changes." In this appendix, we describe the current state of practice in climate change mitigation and adaptation in U.S. urban areas. Our emphasis here is on understanding and characterizing the effectiveness of adaptation practices in particular, because that has thus far proved to be the more challenging task from an urban governance perspective. This discussion provides the background for our consideration in Chapter Two and Appendix B of alternative conceptual frameworks for thinking about adaptation processes and a system of forward-looking indicators in support of adaptive planning and decisionmaking. Looking through the lens of a risk governance conceptual framework, we summarize indicators and evaluation frameworks in Chapter Four that have been developed to track progress on climate change adaptation in the United States and elsewhere.

Given the limited implementation of climate change adaptation in U.S. cities, the literature summarizing the state of practice in U.S. cities is perhaps understandably sparse. We focus as much as possible on practices in U.S. cities but draw examples from around the world as appropriate. The following sections synthesize the available literature on planning that

- describes some of the motivating factors that spur local governments to develop climate change adaptation plans
- characterizes the cities' progress in adaptation planning
- summarizes the most common components of climate change adaptation plans.

Overall, we find that U.S. cities have only recently been giving attention to climate change adaptation, and this has largely been motivated by citizens, advocacy groups, mandates, and weather-related drivers. Mazmanian et al. (2013) offer an intriguing set of hypotheses, collectively called a paradox, about why adaptation has lagged behind mitigation in California, noting that the "think globally, act locally" maxim has been turned on its head. In their assessment, adaptation is more challenging "substantively and politically."

Surveys of adaptation action plans confirm that, until recently, most activities related to climate change focused solely on mitigation (Wheeler, 2008; Aylett, 2014). Nevertheless, there is a growing interest in adaptation planning. Most cities are still only at the awareness-building and assessment phase; few have reached any stage of implementation. In cases where cities have articulated action plans, actions focus largely on capacity building, education, and other "soft" interventions not involving construction activities. Although beyond the scope of this review, the lack of progress from awareness-building and assessment to implementation begs a review of the barriers to adaptation and enablers of organizational change.

Mitigation Initiatives in Urban Areas

The most extensive mitigation efforts under way in the United States are in California, a consequence of the state's two landmark bills: Assembly Bill 32 (State of California, 2006), which sets a target for reductions in the state's emissions to 1990 levels by 2020 and then an 80-percent reduction by 2050; and Senate Bill 375 (State of California, 2008), which requires that MPOs in the state develop sustainable community strategies that demonstrate how the region will reduce GHG emissions through integrated land use and transportation planning. However, there is no requirement that cities and counties adjust local land use plans to reflect these regional plans, nor is state funding contingent on these plans. Instead, the law is built on a suite of incentives through streamlined environmental review.

In a recent survey, the U.S. Conference of Mayors (2014) found that 53 percent of its members said that their cities had committed by official act to reduce GHG emissions. Almost half of all cities reported having conducted a GHG emissions inventory. The U.S. Conference of Mayors survey also asked cities about their preferred means of engaging communities, advancing their energy and climate goals, and the most promising technologies to reduce energy use and GHG emissions. Even with this level of activity, the absence of overarching federal policy and international agreements has discouraged many cities and states from imposing requirements for emissions reductions that would place private enterprises operating in their jurisdictions at a disadvantage relative to competitors operating in places without binding limits.[2]

Motivation for Climate Change Adaptation Planning in Urban Areas

Until recently, adaptive responses to climate change in the United States largely occurred locally because limited federal policies around climate change and the geographic variability of climate change impacts (Cruce, 2009, p. 16). However, U.S. cities are in a period of rapid transition, and much has happened over the last several years. The uptick in activity is likely due to a number of factors, including several major disasters in the last ten years, beginning with Hurricanes Katrina, Gustav, and Ike in the south; tornadoes and flooding in the Midwest; and Hurricane Sandy's toll in the Northeast. The release of the President's Climate Action Plan, the launch of major initiatives of national philanthropic foundations, and growing public support for some action on climate change in New York City, Seattle, and elsewhere have further fueled the increased attention on adaptation activities, most of which have proceeded under the banner of increasing resilience. The Obama administration is moving toward issu-

---

[2]   On June 2, 2014, the U.S. Environmental Protection Agency proposed a Clean Power Plan rule to limit carbon pollution from power plants. This proposed rule represents the most significant step to date toward binding federal limits on GHG emissions (EPA, 2016).

ing guidance to all agencies regarding the incorporation of resilience principles into non–disaster recovery program activities. These resilience guidelines were originally developed to apply to the New York–New Jersey region affected by Hurricane Sandy in 2013 (Finucane et al., 2014).

As summarized in Table D.1, national and transnational governance networks have helped support local climate change responses (Bulkeley, 2010, p. 234). One such national network is the U.S. Conference of Mayors, which developed a U.S. Conference of Mayors Climate Protection Agreement (MCPA) initiative in 2005. Since then, 1,060 mayors from the 50 states, the District of Columbia, and Puerto Rico have signed the MCPA, committing "to reduce emissions in their cities to seven percent below 1990 levels by 2012," consistent with the target reduction outlined in the Kyoto Protocol (U.S. Conference of Mayors, 2015).

Another network of local governments, the International Council for Local Environmental Initiatives, now called ICLEI—Local Governments for Sustainability, currently has 450 member cities and counties across 46 states (ICLEI, 2015). Like the MCPA, ICLEI initially sought to encourage cities to commit to reducing GHG emissions. Surveys of adaptation action plans confirm that most activities related to climate change focused on mitigation (Wheeler, 2008). Nevertheless, there is a growing interest in adaptation planning. ICLEI now describes itself as "the leading global network devoted to local governments engaged in sustainability, climate protection, and clean energy initiatives" and encourages its members to include adaptation in their climate change response (ICLEI, 2015).

While participation in climate change response programs may reflect an initial expression of interest in climate change (or possibly motivate such interest), research indicates that adaptation initiatives are often the result of a combination of place-specific concerns about future climate change impacts and other factors (Ford et al., 2011). One study by Wallis et al. (2011) surveyed regional planning organizations (RPOs), including MPOs and councils of governments, and found that several types of weather events triggered interest in adaptation. A majority of respondents identified increasing frequency of extreme weather events and flooding as factors; approximately half indicated droughts; and a third reported sea-level rise as concerns (Wallis et al., 2011). Ford et al. (2011) conducted a literature review of peer-reviewed publications related to adaptation in developed nations. They also found that experiencing an extreme weather event can induce climate change adaptation efforts.

The literature also suggests several non–weather-related factors:

- political changes, such as the election of a new leader who makes responding to climate change a priority (Ford et al., 2011)
- pressure from citizens for protection from the effects of climate change (Sippel and Jenssen, 2009)

Table D.1
Major Organizations Supporting Local Climate Initiatives

Name	Parent Organization	Purpose
Mayors Climate Protection Agreement (U.S. Conference of Mayors, 2005)	U.S. Conference of Mayors	Participating cities are committed to strive to meet or beat the Kyoto Protocol targets in their own communities; urge their state governments, and the federal government, to enact policies and programs to meet or beat the GHG emission reduction target suggested for the United States in the Kyoto Protocol—7% reduction from 1990 levels by 2012; and urge Congress to pass bipartisan GHG reduction legislation to establish a national emission trading system.
Cities for Climate Protection (ICLEI, 2005)	International Council for Local Environmental Initiatives—Local Governments for Sustainability	The Cities for Climate Protection campaign assists cities in adopting policies and implementing quantifiable measures to reduce local GHG emissions, improve air quality, and enhance urban livability and sustainability.
100 Resilient Cities Initiative	Rockefeller Foundation	This initiative is focused on making cities resilient against large shocks—such as earthquakes, fires, and floods, etc.—and against day-to-day stressors that make urban areas vulnerable to adverse events. Participating cities receive support for developing a resilience strategy and access to private and public sector tools to implement that strategy.
C40 Cities Climate Leadership Group	Bloomberg Philanthropies	C40 is a network of some of the world's largest cities taking action against climate change through reduction of green house gas emissions and climate risks. Policies are centered on measurable and scalable action at the city and local level.
Georgetown Climate Center (2015)	Georgetown University	The nonpartisan Georgetown Climate Center seeks to advance effective climate, energy, and transportation policies in the United States—policies that reduce GHG emissions and help communities adapt to climate change.
Center for Climate and Energy Solutions (2015)	Independent (formerly the Pew Center on Global Climate Change)	The center's purpose is to provide reliable, timely, impartial information and analysis on the scientific, economic, technological, and policy dimensions of climate and energy challenges. The center brings together business, the environmental community, other stakeholders, and policymakers to achieve common understandings and consensus solutions. The center also works closely with policymakers and stakeholders to promote pragmatic, effective policies at the state, national, and international levels.

- local government's desire to have a "green" reputation to attract new businesses and workers (Sippel and Jenssen, 2009)
- the presence of state and federal mandates or initiatives (Wallis, 2011; Tang et al., 2010).

In the study of RPOs mentioned above, Wallis et al. (2011) found that just over half of RPOs confirmed that state or federal initiatives, be they mandatory or voluntary, were motivating factors. Tang et al. (2010) conducted a study analyzing adaptation action plans that were adopted by local governments in the United States to better understand the quality of the plans in terms of the components of climate change awareness, analysis of how climate change would impact the locality, and actions for climate change mitigation and adaptation. Their analysis revealed that the quality of adaptation action plans is significantly influenced by the presence of state mandates, which the study's authors believe serve as a motivation for local governments to develop strong plans.

The role of higher government authorities is complex and may both motivate and hinder effective adaptation planning. Carmin et al.'s (2012, p. 24) survey of ICLEI member cities from around the globe revealed that 6 percent or less of U.S. cities reported that they had "full understanding from their national government," and 36 percent of U.S. cities felt that the "national government does not appreciate the challenges they face and the support they need in adaptation planning." As of 2012, less than a quarter of survey respondents from the United States reported that "their local officials have a strong commitment to adaptation" (Carmin et al., 2012, p. 23). More recent survey data of ICLEI members were not available.

Progress in Adaptation Planning

The intensity of adaptation planning efforts in cities is increasing, although the peer-reviewed literature (Lehmann et al., 2012; Bulkeley et al., 2011; Hunt and Watkiss, 2011; and Moser, 2009 and 2013) lags behind the numerous planning documents emerging directly from cities (see Georgetown Climate Center, 2015, for a full listing). Assessing progress, however, remains challenging.

Wallis et al. (2011, pp. 6–9) describes a simple and now widely adopted four-phase framework to benchmark adaptation progress: awareness, assessment, planning and recommendations, and adaptation and implementation. This framework is one of many possible frameworks; we present others in the discussion of indicators in Appendix A. In the awareness phase, the planning entity begins to think about the impacts of climate change it may face over time. The next phase, assessment, is the period during which the organization doing the planning analyzes how climate change will affect it. During the planning and recommendation phase, the actual adaptation plan is formulated, often using the results of the analysis conducted during the assessment phase. The final phase in the adaptation planning process is adaptation and implementation, during which the adaptation plan developed during the previous phase is adopted and put into action (Wallis et al., 2011).

In 2011, Carmin et al. (2012) surveyed ICLEI members around the globe to better understand the progress being made globally toward climate change adaptation. The survey inquired about the members' activities related to "(1) experience of changing weather and precipitation patterns; (2) risk and vulnerability assessment;

(3) planning activities; (4) support for and influences on planning; (5) challenges and benefits; and (6) location characteristics." The study found that although nearly 60 percent of U.S. cities were actively involved in some form of adaptation planning, less than a third were engaged in or had completed a climate risk assessment. Moreover, the rate of climate risk assessment among U.S. ICLEI members was the lowest among all the responding regions (Africa, Asia, Australia and New Zealand, Canada, Europe, Latin America, and the United States).

Wallis et al. (2011) reported similar findings from their survey of U.S. RPOs. 96.6 percent of respondents indicated involvement in climate adaptation planning at the *awareness* phase, but only 29.2 percent indicated that they had moved beyond the awareness phase into *assessment*, and only 27.0 percent had reached the *planning* phase. Moreover, only 8.9 percent of respondents indicated reaching the final phase of *adoption and implementation* (Wallis et al., 2011). More recently, the U.S. Conference of Mayors reported that 40 percent of surveyed cities were developing a climate adaptation plan, and the same percentage of cities were also engaged in assessing and creating responses to predicted climate impacts (U.S. Conference of Mayors, 2014).[3] The variability in results among these studies could be due to different survey pools, wording of questions, or definition of phases. Nevertheless, these studies show that there is a growing awareness of the need for climate adaptation planning, though moving beyond the awareness and initial plan development phases proves to be challenging.

Consistent with the above-mentioned studies, the Georgetown Climate Center developed an online interactive tool with a listing of state and local adaptation plans, state policies, and state agency plans (Georgetown Climate Center, 2015). The tool enables a user to view the progress of plan implementation marked by the number of goals achieved or in progress by sector for state-led plans: 15 states have finalized state-level adaptation plans, eight states and the District of Columbia have state-led planning efforts under way, and 27 states have one or more finalized local adaptation plans. A few states still do not have activities under way as of this writing (September 2016). California and Massachusetts have over 300 discrete goals in each state-level plan. Among the goals completed to date, most fall within planning and capacity building, similar to the phases of awareness, assessment, and planning in Wallis et al.'s 2011 framework.

Conversely, since mitigation-planning efforts have been under way in many cities for a decade or more, mitigation actions have produced measurable reductions in GHG emissions in several areas. As a complement to Carmin et al.'s 2012 adaptation-focused survey of ICLEI members, Aylett's 2014 survey focuses more on the current state of mitigation actions among ICLEI members. When asked about their cities' focus on

---

[3] The U.S. Conference of Mayors (2014) conducted an electronic survey on climate change mitigation and adaptation actions between November 25, 2013, and January 14, 2014. Survey responses totaled 282, most representing populations of 30,000 or more.

climate adaptation and mitigation, 58 percent of U.S. respondents reported engaging in both adaptation and mitigation; 41 percent reported focusing only on mitigation. These results support the trend of increasing interest in adaptation planning previously discussed but also highlight that there is still significant focus only on mitigation actions in many U.S. cities. According to Aylett's survey, 98 percent of U.S. city respondents have completed a GHG emissions inventory, the first step in climate change mitigation planning, and reported that the top areas for measurable GHG reductions from mitigation efforts were in local government buildings, local government fleets, residential energy use, and waste reduction.

To better understand how cities are incorporating climate change planning into government planning processes, Aylett (2014) asked a series of questions about where plans were documented in relation to other planning documents. Many cities reported creating specific plans for mitigation or adaptation, as opposed to including actions in long-range development plans, city master plans, or other relevant spatial development plans. Among cities with specific plans, a global comparison of inaugural climate change plans to current plans shows a preference toward integrating adaption planning with mitigation planning; however, mitigation-specific plans still outnumber integrated plans in the United States (Aylett, 2014).[4]

Several barriers have been identified as obstacles to motivating the different stages of adaptation planning in the United States, including how to handle uncertainty in climate change projections in a policy setting, lack of financial and human capital to begin and sustain actions, fragmented decisionmaking, institutional constraints, insufficient political leadership, and variations in risk perceptions and cultural values (Bierbaum et al., 2014; Aylett, 2014; Mimura et al., 2014). One way that local governments are trying to overcome these barriers is leveraging partnerships with other cities, community groups, NGOs, and business. Several studies indicate that local governments are engaging in partnerships to help design and implement climate change responses (Carmin et al., 2012; Wallis et al., 2011; Aylett, 2014). In Carmin et al.'s 2012 survey, U.S. respondents cited partnerships with other cities most commonly, followed by partnerships with community groups (35 percent and 25 percent, respectively). Similarly, Wallis et al. (2011) found that nearly 62 percent of RPOs were working with other organizations, including colleges, universities, state agencies, and federal agencies.

Research shows that belonging to a global climate action organization has not necessarily resulted in climate change mitigation or adaptation actions (Krause, 2012; Reams et al., 2012). A survey of the implementation status of climate, energy, and transportation mitigation policies in 329 cities revealed that participation in ICLEI's Cities for Climate Protection program had only "a small to moderate impact on the local

---

[4]   Among U.S. respondents, 47 percent cited specific mitigation plans, compared with 34 percent globally, and 25 percent cited integrated mitigation and adaptation plans, compared with 39 percent globally. Most commonly, cities reported incorporating mitigation or adaptation actions into some other local government plan, such as long-range development plans.

implementation of GHG-relevant actions," but the MCPA had no such effect (Krause, 2012). Further, in a study of just ICLEI members, Reams et al. (2012) assessed the ICLEI milestone completion for 257 cities across 40 states and found that 110 of the cities had not completed any of the milestones, and 36 had completed only one.[5] While 58 cities had achieved the third milestone (developing a local climate-action plan), only 31 of the cities had implemented the plans they developed. In their analysis, Reams et al. (2012) found that the most significant predictor of milestone completion was length of membership. These findings suggest that low rates of completion of milestones may be correlated with the time it takes to complete each step of the process.

Similarly, Sharp, Daley, and Lynch (2011) found that the duration of ICLEI membership was positively correlated with milestone completion; however, their study revealed other contributing factors.[6] The authors found that governance form (mayor versus city manager model), budgets, economic conditions, interest group influence, and level of organization of nonprofit environmental groups can all shape the level of commitment and sustainment of focus on climate-related initiatives. For example, in mayoral cities, fiscal stress and a high ratio of manufacturing industry to other types of industry were correlated with fewer completed milestones. Conversely, greater numbers of environmental nonprofits and citizens with bachelor's degrees or higher were positively correlated with milestone completion in mayoral cities. In cities with managers, metropolitan fragmentation, defined as "the number of general purpose municipal governments and counties existing in a study city's metropolitan area," was negatively correlated with completion of milestones (Sharp, Daley, and Lynch, 2011). These results offer insight into some relevant factors that may influence progression from assessment to plan implementation and evaluation.

Other campaigns and initiatives, such as the Rockefeller Foundation's 100 Resilient Cities program and the U.S. Department of Housing and Urban Development's (HUD's) National Disaster Resilience Competition (NDRC), are focused on providing resources and tools, rather than just frameworks, to help cities move past awareness and assessment phases into implementation and evaluation (Rockefeller Foundation, 2015; HUD, 2014b). These programs are focused on incorporating climate change mitigation and adaptation actions into the broader concept of citywide resilience. That is, these initiatives recognize that climate change responses are an integral component of resilience but also recognize that there are other non–climate-related stressors that contribute to city vulnerabilities (Rockefeller Foundation, 2015; HUD, 2014b). For example, the 100 Resilient Cities program views resilience through four dimensions:

---

[5] ICLEI milestones are (1) conduct a local GHG emission inventory, (2) adopt emission-reduction targets, (3) develop a local climate-action plan, (4) implement the plan, and (5) monitor and verify the results.

[6] Sharp, Daley, and Lynch (2011) surveyed a random sample of 122 U.S. cities with populations of at least 100,000. They were interested in identifying the factors that influenced the decision to become an ICLEI member and factors that influenced milestone completion for member cities.

(1) leadership and strategy, (2) health and well-being, (3) economy and society, and (4) infrastructure and environment. The 100 Resilient Cities program has already identified 67 cities (globally) to receive financial and logistic support in hiring a chief resilience officer and developing a resilience plan, access to public and private sector tools to implement the strategy, and membership in a global network of resilience practitioners (Rockefeller Foundation, 2015). Because this program is still young, the impact of this initiative cannot yet be fully measured; however, the program is attempting to address barriers to implementation through practical solutions.[7]

Similarly, HUD's NDRC will fund local resilience projects for communities that have experienced a natural disaster and will provide technical assistance and training workshops to eligible state and local governments (HUD, 2014a). This competition was developed in response to demand from state, local, and tribal leaders for assistance in helping make their communities safer, more secure, and more resilient against future disasters (HUD, 2014b). While this competition is based on weather-related climate change impacts and rebuilding after natural disasters, the intent is that taking these actions will help protect community assets, people, and economic activity by making the whole community more resilient to future disasters. This competition is currently in phase II of the application process and expects to announce funding in December 2015 (HUD, 2014a).

Common Components of Adaptation Plans

Given the nascent nature of climate change adaptation–focused plans in most cities, adaptation activities prescribed within the plans often emphasize planning and education activities, rather than actions with measurable outcomes related to risk reduction or improved resilience. Carmin et al. (2012) found that the most common adaptation activities being undertaken were "(1) meeting with local government departments on adaptation; (2) searching the web or literature for information on adaptation; (3) forming a commission or task force to support adaptation planning; and (4) developing partnerships with NGOs, other cities, businesses, or community groups." As first steps in the planning process, these actions assist with capacity building and increasing institutional literacy or awareness. Similarly, other scholars found that 90 percent of the publications reviewed involved such interventions as "developing management strategies, plans, policies, regulations, guidelines, or operating networks to guide current and/or future planning" (Ford et al., 2011).

In an assessment of more advanced city adaptation plans, Tang et al. (2010) found that "current local climate change action plans focus predominantly on the built envi-

---

[7]   The first 32 cities in the 100 Resilient Cities program were announced in December of 2013; 35 more cities were announced in December 2014, and the last challenge for remaining spots will remain open to applicants late in 2015 (Rockefeller Foundation, 2015).

ronment (e.g. energy, transportation, wastes and buildings)." These findings are also consistent with Wallis et al.'s 2011 findings that the most common recommendations RPOs reported working on were "sustainability and/or smart growth planning" and "encouraging higher density development." RPOs further indicated that "outreach and education," "infrastructure development," and "recommendations based on best practices" were the tools they utilized to implement their plans (Wallis et al., 2011, p. 9).

Bibliography

ACC—*see* America's Climate Choices.

America's Climate Choices, *Informing an Effective Response to Climate Change*, 2010.

Arctic Council, "Glossary of Terms," *Arctic Resilience Interim Report 2013*, Stockholm, Sweden: Stockholm Environment Institute and Stockholm Resilience Centre, 2013.

Argyris, C., and D. Schön, *Organizational Learning: A Theory of Action Perspective*, Reading, Mass.: Addison Wesley, 1978.

Arup, *City Resilience Framework: City Resilience Index*, New York: The Rockefeller Foundation, 2014.

Asian Development Bank, *An Introduction to Results Management: Principles, Implications, and Applications*, technical report, 2006.

Association of Bay Area Governments and the Metropolitan Transportation Commission, *Plan Bay Area: Strategy for a Sustainable Region*, Final Performance Assessment Report, July 2013.

Australian Agency for International Development, "AusGUIDElines: 1. The Logical Framework Approach," *AusGUIDE*, 2003.

Aylett, Alexander, *Progress and Challenges in the Urban Governance of Climate Change: Results of a Global Survey*, Cambridge, Mass.: MIT, 2014.

Baer, W. C., "General Plan Evaluation Criteria: An Approach to Making Better Plans," *Journal of the American Planning Association*, Vol. 63, No. 3, 1997, pp. 329–344.

Baker, Ingrid, Ann Peterson, Greg Brown, and Clive McAlpine, "Local Government Response to the Impacts of Climate Change: An Evaluation of Local Climate Adaptation Plans," *Landscape and Urban Planning*, Vol. 107, No. 2, 2012, pp. 127–136.

Bastoe, Per O., and Clarence Henderson, *An Introduction to Results Management: Principles, Implications, and Applications*, Metro Manila, Philippines: Asian Development Bank, 2006.

Berke, P., and D. Godschalk, "Searching for the Good Plan: A Meta-Analysis of Plan Quality Studies," *Journal of Planning Literature*, Vol. 23, No. 3, 2009, pp. 227–240.

Bierbaum, Rosina, Joel B. Smith, Arthur Lee, Maria Blair, Lynne Carter, F. Stuart Chapin, and Laura Verduzco, "A Comprehensive Review of Climate Adaptation in the United States: More Than Before, But Less Than Needed," *Mitigation and Adaptation Strategies for Global Change*, Vol. 18, No. 3, 2012, pp. 361–406.

Bierbaum, R., A. Lee, J. Smith, M. Blair, L. M. Carter, F. S. Chapin III, P. Fleming, S. Ruffo, S. McNeeley, M. Stults, L. Verduzco, and E. Seyller, "Chapter 28: Adaptation," in J. M. Melillo, Terese (T.C.) Richmond, and G. W. Yohe, eds., *Climate Change Impacts in the United States: The Third National Climate Assessment*, U.S. Global Change Research Program, 2014, pp. 670–706.

Biesbroek, G. R., J. E. M. Klostermann, C. J. A. M. Termeer, and P. Kabat, "On the Nature of Barriers to Climate Change Adaptation," *Regional Environmental Change*, Vol. 13, No. 5, 2013, pp. 1119–1129.

Birkmann, J., O. D. Cardona, M. L. Carreño, A. H. Barbat, M. Pelling, S. Schneiderbauer, S. Kienberger, M. Keiler, D. Alexander, and P. Zeil, "Framing Vulnerability, Risk and Societal Responses: The MOVE Framework," *Natural Hazards*, Vol. 67, No. 2, 2013, pp. 193–211.

Bloom, Evan, *Changing Midstream: Providing Decision Support for Adaptive Strategies Using Robust Decision Making: Applications in the Colorado River Basin*, Santa Monica, Calif.: RAND Corporation, RGSD-348, 2015. As of October 16, 2016: http://www.rand.org/pubs/rgs_dissertations/RGSD348.html

Bloomberg, Michael R., Jeffrey D. Sachs, and Gillian M. Small, "Climate Change Adaptation in New York City: Building a Risk Management Response," *Annals of the New York Academy of Sciences*, Vol. 1196, No. 1, 2010, pp. 1–3.

Bloomberg Philanthropies, "Environment—Bloomberg Philanthropies," 2015. As of April 27, 2015: http://www.bloomberg.org/program/environment/#sustainable-cities

Bosetti, V., C. Carraro, E. Massetti, and M. Tavoni, eds., *Climate Change Mitigation, Technological Innovation and Adaptation: A New Perspective on Climate Policy*, FEEM, 2013.

Bours, Dennis, Colleen McGinn, and Patrick Pringle, *Monitoring and Evaluation for Climate Change Adaptation: A Synthesis of Tools, Frameworks and Approaches*, Phnom Penh, Cambodia, and Oxford, UK: SEA Change Community of Practice and UKCIP, 2013. As of October 16, 2016: http://www.seachangecop.org/node/2588

Brooks, Nick, Simon Anderson, Jessica Ayers, Ian Burton, and Ian Tellam, "Tracking Adaptation and Measuring Development," IIED Working Paper No. 1, London and Edinburgh: IIED, 2011.

Brooks, Nick, Simon Anderson, Ian Burton, Susannah Fisher, Neha Rai, and Ian Tellam, *An Operational Framework for Tracking Adaptation and Measuring Development (TAMD)*, 2013.

Brown, C., "The End of Reliability," editorial, *Journal of Water Resources Planning and Management*, Vol. 136, No. 2, 2010, pp. 143–145.

Brown, C., and R. Wilby, "An Alternate Approach to Assessing Climate Risks," *EOS, Transactions American Geophysical Union*, Vol. 93, No. 41, 2012, p. 401.

Bulkeley, Harriet, "Cities and the Governing of Climate Change," *Annual Review of Environment and Resources*, Vol. 35, No. 1, 2010, pp. 229–253.

Bulkeley, Harriet, Heike Schroeder, Katy Janda, Jimin Zhao, Andrea Armstrong, Shu Yi Chu, and Shibani Ghosh, "The Role of Institutions, Governance, and Urban Planning for Mitigation and Adaptation," in D. Hoornweg, M. Freire, M. J. Lee, P. Bhada-Tata, and B. Yuen, eds., *Cities and Climate Change: Responding to an Urgent Agenda*, World Bank, 2011, pp. 125–129.

Carmin, JoAnn, Nikhil Nadkarni, and Christopher Rhie, "Progress and Challenges in Urban Climate Adaptation Planning: Results of a Global Survey," Cambridge, Mass.: MIT, 2012.

Carter, T. R., R. N. Jones, S. B. X. Lu, C. Conde, L. O. Mearns, B. C. O'Neill, M. D. A. Rounsevell, and M. B. Zurek, "New Assessment Methods and the Characterisation of Future Conditions," in M. L. Parry, O. F. Canziani, J. P. Palutikof, P. J. v. d. Linden, and C. E. Hanson, *Climate Change 2007: Impacts, Adaptation and Vulnerability; Contribution of Working Group II to the Fourth Assessment Report of the Intergovernmental Panel on Climate Change*, Cambridge, UK: Cambridge University Press, Vol. 1, 2007, pp. 33–171.

Center for Climate and Energy Solutions, "About," 2015. As of October 16, 2016: http://www.c2es.org/about

Centers for Disease Control and Prevention, Program Performance and Evaluation Office, "Introduction to Program Evaluation for Public Health Programs: A Self-Study Guide," 2012. As of November 3, 2015:
http://www.cdc.gov/eval/guide/glossary/

Centre for Environmental Risks and Futures, *Evaluating the Risk Assessment of Adaptation Reports Under the Adaptation Reporting Power: Final Summary*, Cranfield University, UK, 2012.

Chandra, Anita, Joie Acosta, Stefanie Howard, Lori Uscher-Pines, Malcolm Williams, Douglas Yeung, Jeffrey Garnett, and Lisa S. Meredith, *Building Community Resilience: A Way Forward to Enhance National Health Security*, Santa Monica, Calif.: RAND Corporation, TR-915-DHHS, 2011. As of October 16, 2016:
http://www.rand.org/pubs/technical_reports/TR915.html

Chandra, A., M. Williams, A. Plough, et al., "Getting Actionable About Community Resilience: Los Angeles County, Community Disaster Resilience Project," *American Journal of Public Health*, July 2013.

City of Los Angeles, Department of Public Works, Bureau of Sanitation, "One Water LA 2040," 2014. As of October 16, 2016:
http://www.lacitysan.org/irp/OW_Documents/OneWaterLA_Fact_Sheet_2014.pdf

Coastal Protection and Restoration Authority of Louisiana, *Louisiana's Comprehensive Master Plan for a Sustainable Coast*, Baton Rouge, La., 2012. As of April 30, 2012:
http://www.coastalmasterplan.louisiana.gov/2012-master-plan/final-master-plan/

Coastal Protection and Restoration Authority, "About," 2015. As of May 29, 2015:
http://coastal.la.gov/about/history/

Cox, J., and Anthony Louis, "Confronting Deep Uncertainties in Risk Assessment," *Risk Analysis*, Vol. 32, No. 10, 2012, pp. 1607–1629.

Cronin, M. A., C. Gonzalez, and J. D. Sterman, "Why Don't Well-Educated Adults Understand Accumulation? A Challenge to Researchers, Educators, and Citizens," *Organizational Behavior and Human Decision Processes*, Vol. 108, 2009, pp. 116–130.

Cruce, Terri L., "Adaptation Planning—What U.S. States and Localities Are Doing," Pew Center on Global Climate Change, 2009.

da Silva, J., and B. Morera, *City Resilience Framework: City Resilience Index*, Rockefeller Foundation, Arup, 2014.

DeCanio, S. J., and A. Fremstad, "Game Theory and Climate Diplomacy," *Ecological Economics*, Vol. 85, 2013, pp. 177–187.

Department for Communities and Local Government and the Rt Hon Sir Eric Pickles, "Councils' Red Tape Cut as 4,700 Whitehall Targets Slashed," press release, October 2010. As of April 30, 2014:
https://www.gov.uk/government/news/councils-red-tape-cut-as-4-700-whitehall-targets-slashed

Department for Environment, Food and Rural Affairs, *Adaptation Reporting Power: Received Reports*, London: UK Government, 2012. As of October 16, 2016:
https://www.gov.uk/government/publications/adaptation-reporting-power-received-reports

Department for International Development, "About Us," undated. As of May 12, 2014:
https://www.gov.uk/government/organisations/department-for-international-development/about

Dessai, S., and M. Hulme, "Assessing the Robustness of Adaptation Decisions to Climate Change Uncertainties: A Case Study on Water Resources Management in the East of England," *Global Environmental Change*, Vol. 17, No. 1, 2007, pp. 59–72.

Dewar, J. A., *Assumption-Based Planning—A Tool for Reducing Avoidable Surprises*, Cambridge, UK: Cambridge University Press, 2002.

Dewar, James A., Carl H. Builder, William M. Hix, and Morlie H. Levin, *Assumption-Based Planning: A Planning Tool for Very Uncertain Times*, Santa Monica, Calif.: RAND Corporation, MR-114-A, 1993. As of October 16, 2016:
http://www.rand.org/pubs/monograph_reports/MR114.html

Dickson, Eric, Judy L. Baker, Daniel Hoornweg, and Asmita Tiwari, *Urban Risk Assessments: Understanding Disaster and Climate Risk in Cities*, Urban Development Series, Washington, D.C.: World Bank, 2012. As of October 16, 2016:
http://documents.worldbank.org/curated/en/2012/06/16499064/
urban-risk-assessments-understanding-disaster-climate-risk-cities

Diener, E., *Well-Being for Public Policy*, New York: Oxford University Press, 2009.

Dodge, R., A. Daly, J. Huyton, and L. Sanders, "The Challenge of Defining Wellbeing," *International Journal of Wellbeing*, Vol. 2, No. 3, 2012, pp. 222–235.

Drew, G. H., S. J. T. Pollard, S. A. Rocks, and S. R. Jude, *Evaluating the Risk Assessments of Reporting Authorities Under the Climate Change Act, 2008*, Cranfield University, UK: The Collaborative Centre of Excellence for Understanding and Managing Natural and Environmental Risks (Risk Centre), 2010.

Eisenack, K., and R. Stecker, "An Action Theory of Adaptation to Climate Change," Earth System Governance Working Paper No. 13, Lund and Amsterdam: Earth System Governance Project, 2011.

Engle, N. L., A. de Bremond, E. L. Malone, and R. H. Moss, "Towards a Resilience Indicator Framework for Making Climate-Change Adaptation Decisions," *Mitigation and Adaptation Strategies for Global Change*, Vol. 19, No. 8, 2014, pp. 1295–1312.

Environmental Protection Agency, "Fact Sheet: Clean Power Plan Overview," April 11, 2016. As of October 16, 2016:
http://www2.epa.gov/carbon-pollution-standards/fact-sheet-clean-power-plan-overview

EPA—*see* Environmental Protection Agency.

Finucane, Melissa L., Noreen Clancy, Henry H. Willis, and Debra Knopman, *The Hurricane Sandy Rebuilding Task Force's Infrastructure Resilience Guidelines: An Initial Assessment of Implementation by Federal Agencies*, Santa Monica, Calif.: RAND Corporation, RR-841-DHS, 2014. As of October 16, 2016:
http://www.rand.org/pubs/research_reports/RR841.html

Fischbach, Jordan R., Robert J. Lempert, Edmundo Molina-Perez, Abdul Ahad Tariq, Melissa L. Finucane, and Frauke Hoss, *Managing Water Quality in the Face of Uncertainty: A Robust Decision Making Demonstration for EPA's National Water Program*, Santa Monica, Calif.: RAND Corporation, RR-720-EPA, 2015. As of November 29, 2016:
http://www.rand.org/pubs/research_reports/RR720.html

Fischbach, Jordan R., David R. Johnson, David S. Ortiz, Benjamin P. Bryant, Matthew Hoover, and Jordan Ostwald, *Coastal Louisiana Risk Assessment Model: Technical Description and 2012 Coastal Master Plan Analysis Results*, Santa Monica, Calif.: RAND Corporation, TR-1259-CPRA, 2012. As of October 16, 2016:
http://www.rand.org/pubs/technical_reports/TR1259.html

Finnish Ministry of Agriculture and Forestry, *Evaluation of the Implementation of Finland's National Strategy for Adaptation to Climate Change*, 2009. As of October 16, 2016:
http://mmm.fi/documents/1410837/1721034/Adaptation_Strategy_evaluation.pdf/
043c0964-58c5-4fce-8924-cc47748cf766

Fixsen, D. L., S. F. Naoom, K. A. Blase, and R. M. Friedman, *Implementation Research: A Synthesis of the Literature*, Tampa, Fla.: Florida Mental Health Institute, National Implementation Research Network, FMHI Publication #231, 2005. As of October 16, 2016: www.fpg.unc.edu/~nirn/resources/publications/Monograph/

Ford, James D., Lea Berrang-Ford, Alex Lesnikowski, Magda Barrera, and S. Jody Heymann, "How to Track Adaptation to Climate Change: A Typology of Approaches for National-Level Application," *Ecology and Society*, Vol. 18, No. 3, 2013, p. 40.

Ford, J. D., L. Berrang-Ford, and J. Paterson, "A Systematic Review of Observed Climate Change Adaptation in Developed Nations: A Letter," *Climatic Change*, Vol. 106, No. 2, 2011, pp. 327–336.

Ford, J. D., and D. King, "A Framework for Examining Adaptation Readiness," *Mitigation and Adaptation Strategies for Global Change*, 2013, pp. 1–22.

Frankel-Reed, Jennifer, Nick Brooks, Pradeep Kurukulasuriya, and Bo Lim, "A Framework for Evaluating Adaptation to Climate Change: Evaluating Climate Change and Development," *Evaluating Climate Change and Development*, Vol. 8, 2011, p. 285.

Funke, M., and M. Paetz, "Environmental Policy Under Model Uncertainty: A Robust Optimal Control Approach," *Climatic Change*, Vol. 107, 2011, pp. 225–239.

Füssel, Hans-Martin, "Assessing Adaptation to the Health Risks of Climate Change: What Guidance Can Existing Frameworks Provide?" *International Journal of Environmental Health Research*, Vol. 18, No. 1, 2008, pp. 37–63.

Füssel, H.-M., "Vulnerability: A Generally Applicable Conceptual Framework for Climate Change Research," *Global Environmental Change*, Vol. 17, No. 2, 2007, pp. 155–167.

Füssel, H.-M., and Richard J. T. Klein, "Climate Change Vulnerability Assessments: An Evolution of Conceptual Thinking," *Climatic Change*, Vol. 75, No. 3, 2006, pp. 301–329.

Gagnon-Lebrun, Frédéric, and Shardul Agrawala, "Implementing Adaptation in Developed Countries: An Analysis of Progress and Trends," *Climate Policy*, Vol. 7, No. 5, 2007, pp. 392–408.

Georgetown Climate Center, "State and Local Adaptation Plans," 2016. As of May 3, 2016: http://www.georgetownclimate.org/adaptation/state-and-local-plans

Greenfield, Victoria A., Valerie L. Williams, and Elisa Eiscman, *Using Logic Models for Strategic Planning and Evaluation: Application to the National Center for Injury Prevention and Control*, Santa Monica, Calif.: RAND Corporation, TR-370-NCIPC, 2006. As of October 16, 2016: http://www.rand.org/pubs/technical_reports/TR370.html

Groves, D. G., E. W. Bloom, R. J. Lempert, J. R. Fischbach, J. Nevills, and B. Goshi, "Developing Key Indicators for Adaptive Water Planning," *Journal of Water Resources Planning and Management*, Vol. 141, No. 7, September 15, 2014.

Groves, David G., Robert J. Lempert, Debra Knopman, and Sandra H. Berry, "Preparing for an Uncertain Climate in the Inland Empire: Identifying Robust Water-Management Strategies," Santa Monica, Calif.: RAND Corporation, DB-550-NSF, 2008. As of October 16, 2016: http://www.rand.org/pubs/documented_briefings/DB550.html

Groves, D. G., and C. Sharon, "Planning Tool to Support Planning the Future of Coastal Louisiana," *Journal of Coastal Research*, Vol. 67, 2013, pp. 147–161.

Groves, David G., Christopher Sharon, and Debra Knopman, *Planning Tool to Support Louisiana's Decisionmaking on Coastal Protection and Restoration: Technical Description*, Santa Monica, Calif.: RAND Corporation, TR-1266-CPRA, 2012. As of October 16, 2016: http://www.rand.org/pubs/technical_reports/TR1266.html

Groves, David G., Jordan R. Fischbach, Evan Bloom, Debra Knopman, and Ryan Keefe, *Adapting to a Changing Colorado River: Making Future Water Deliveries More Reliable Through Robust Management Strategies*, Santa Monica, Calif.: RAND Corporation, RR-242-BOR, 2013. As of October 16, 2016:
http://www.rand.org/pubs/research_reports/RR242.html

Groves, David G., Jordan R. Fischbach, Debra Knopman, David R. Johnson, and Kate Giglio. *Strengthening Coastal Planning: How Coastal Regions Could Benefit from Louisiana's Planning and Analysis Framework*, Santa Monica, Calif.: RAND Corporation, RR-437-RC, 2014. As of October 16, 2016:
http://www.rand.org/pubs/research_reports/RR437.html

Gunderson, L., and C. S. Holling, *Panarchy: Understanding Transformations in Human and Natural Systems,* Washington, D.C.: Island Press, 2002.

Haasnoot, M., J. H. Kwakkel, W. E. Walker, and J. ter Maat, "Dynamic Adaptive Policy Pathways: A Method for Crafting Robust Decisions for a Deeply Uncertain World," *Global Environmental Change*, Vol. 23, No. 2, 2013, pp. 485–498.

Hallegatte, S., A. Shah, R. Lempert, C. Brown, and S. Gill, *Investment Decision Making Under Deep Uncertainty: Application to Climate Change*, Washington, D.C.: World Bank, 2012.

Halsnæs, K., P. Shukla, D. Ahuja, G. Akumu, R. Beale, J. Edmonds, C. Gollier, A. Grübler, M. Ha Duong, A. Markandya, M. McFarland, E. Nikitina, T. Sugiyama, A. Villavicencio, and J. Zou, "Framing Issues," in B. Metz, O. R. Davidson, P. R. Bosch, R. Dave, and L. A. Meyer, eds., *Climate Change 2007: Mitigation; Contribution of Working Group III to the Fourth Assessment Report of the Intergovernmental Panel on Climate Change*, Cambridge, United Kingdom, and New York: Cambridge University Press, 2007.

Hargrove, E., "Presidential Leadership: Skill in Context," *Politics and Policy*, Vol. 30, No. 2, 2002.

Hazards and Vulnerability Research Institute, "Social Vulnerability Index (SoVI) 2006-10," Department of Geography, University of South Carolina, last updated October 30, 2013. As of October 16, 2016:
http://webra.cas.sc.edu/hvri/products/sovi.aspx

Helm, C., and D. F. Sprinz, "Measuring the Effectiveness of International Environmental Regimes," *Journal of Conflict Resolution*, Vol. 45, No. 5, 2000, pp. 630–652.

Hennlock, M., "Robust Control in Global Warming Management: An Analytical Dynamic Integrated Assessment," Washington, D.C.: Resources for the Future, 2009.

Hinkel, J., "Indicators of Vulnerability and Adaptive Capacity: Towards a Clarification of the Science–Policy Interface," *Global Environmental Change*, Vol. 21, No. 1, 2011, pp. 198–208.

Hirsch, M., "Game Theory and International Environmental Cooperation," *Journal of Energy & Natural Resources Law*, Vol. 27, 2009, p. 503.

Holling, C. S., *Resilience and Stability of Ecological Systems, Annual Review of Ecology and Systematics*, Vol. 4, 1973, pp. 1–23.

Holling, C. S., *Adaptive Environmental Assessment and Management*, Chichester, UK: John Wiley and Sons, 1978.

Holling, C. S., "Engineering Resilience Versus Ecological Resilience," *Engineering Within Ecological Constraints*, 1966, pp. 31–44.

Hovi, J., D. F. Sprinz, and A. Underdal, "The Oslo-Potsdam Solution to Measuring Regime Effectiveness: Critique, Response, and the Road Ahead," *Global Environmental Politics*, Vol. 3, No. 3, August 2003, pp. 74–96.

HUD—*see* U.S. Department of Housing and Urban Development.

Hunt, Alistair, and Paul Watkiss, "Climate Change Impacts and Adaptation in Cities: A Review of the Literature," *Climatic Change*, Vol. 104, No. 1, 2011, pp. 13–49.

ICLEI—*see* International Council for Local Environmental Initiatives.

IIED—*see* International Institute for Environment and Development.

Intergovernmental Panel on Climate Change, *Managing the Risks of Extreme Events and Disasters to Advance Climate Change Adaptation: A Special Report of Working Groups I and II of the Intergovernmental Panel on Climate Change*, Cambridge, UK, and New York: Cambridge University Press, 2012.

Intergovernmental Panel on Climate Change, "Annex III: Glossary," in *Climate Change 2013: The Physical Science Basis. Contribution of Working Group I to the Fifth Assessment Report of the Intergovernmental Panel on Climate Change*, Cambridge, UK, and New York: Cambridge University Press, 2013.

Intergovernmental Panel on Climate Change, "Summary for Policymakers," in *Climate Change 2014: Impacts, Adaptation, and Vulnerability. Part A: Global and Sectoral Aspects. Contribution of Working Group II to the Fifth Assessment Report of the Intergovernmental Panel on Climate Change*, Cambridge, UK, and New York: Cambridge University Press, 2014a.

Intergovernmental Panel on Climate Change, *Climate Change 2014: Impacts, Adaptation, and Vulnerability. Part A: Global and Sectoral Aspects. Contribution of Working Group II to the Fifth Assessment Report of the Intergovernmental Panel on Climate Change*, Cambridge, UK, and New York: Cambridge University Press, 2014b.

Intergovernmental Panel on Climate Change, "Annex II: Glossary," in *Climate Change 2014: Synthesis Report. Contribution of Working Groups I, II and III to the Fifth Assessment Report of the Intergovernmental Panel on Climate Change*, Geneva, Switzerland, 2014c, pp. 117–130. As of November 3, 1016:
https://www.ipcc.ch/pdf/assessment-report/ar5/syr/SYR_AR5_FINAL_full_wcover.pdf

International Council for Local Environmental Initiatives, "Fact Sheet," 1993. As of October 16, 2016:
http://www.abag.ca.gov/lgep/pdfs/climate/ICLEI_Fact-Sheet.pdf

International Council for Local Environmental Initiatives, "FAQ: About ICLEI—Local Governments for Sustainability," 2016a. As of May 3, 2016:
http://www.iclei.org/about/who-is-iclei/faq.html

International Council for Local Environmental Initiatives, "USA Members," 2016b. As of May 3, 2016:
http://www.iclei.org/iclei-members/iclei-members.html

International Institute for Environment and Development, "Tracking Adaptation and Measuring Development (TAMD)," in *Ghana, Kenya, Mozambique, Nepal, Pakistan: Summary Report*, London, May 2013.

International Risk Governance Council, "White Paper on Risk Governance: Towards an Integrative Approach," 2006. As of May 3, 2016:
https://www.irgc.org/risk-governance/irgc-risk-governance-framework/

International Standards Organization, "ISO 14001," 2015. As of October 16, 2016:
https://www.iso.org/obp/ui/#iso:std:iso:14001:ed-3:v1:en

Ionescu, C., Richard J. T. Klein, Jochen Hinkel, K. S. Kavi Kumar, and Rupert Klein, "Towards a Formal Framework of Vulnerability to Climate Change," *Environmental Modeling & Assessment*, Vol. 14, No. 1, 2009, pp. 1–16.

IPCC—*see* Intergovernmental Panel on Climate Change.

Isley, S., R. Lempert, S. Popper, and R. Vardavas, "The Effect of Near-Term Policy Choices on Long-Term Greenhouse Gas Transformation Pathways," *Global Environment Change*, Vol. 34, September 2015, pp. 147–158.

Jabareen, Y. R., "Building a Conceptual Framework: Philosophy, Definitions, and Procedure," *International Journal of Qualitative Methods*, Vol. 8, No. 4, 2009, pp. 49–62.

Jones, R. N., A. Patwardhan, S. Cohen, S. Dessai, A. Lammel, R. Lempert, M. M. Q. Mirza, and H. v. Storch, "Chapter 2: Foundations for Decision Making," *Climate Change 2014: Impacts, Adaptation, and Vulnerability*, Intergovernmental Panel on Climate Change, 2014.

Kahan, D. M., and D. Braman, "Cultural Cognition and Public Policy," *Yale Law and Policy Review*, Vol. 24, 2006, pp. 147–170.

Kahan, D. M., K. Carpenter, and M. Berger, *Proselytizing Normality: An Experimental Test*, Evidence-Based Science Communication Initiative Report No. 2, Cultural Cognition Project at Yale Law School, Ver. 1.1, November 14, 2014.

Kalra, N., S. Hallegatte, R. Lempert, C. Brown, A. Fozzard, S. Gill, and A. Shah, "Agreeing on Robust Decisions: A New Process for Decision Making Under Deep Uncertainty," Policy Research Working Paper, World Bank, 2014.

Keen, M., V. A. Brown, and R. Dyball, "Social Learning: A New Approach to Environmental Management," in M. Keen, V. A. Brown, and R. Dyball, eds., *Social Learning in Environmental Management: Towards a Sustainable Future*, London and New York: Earthscan, 2005, pp. 2–59.

Keeney, R. L., and H. Raiffa, *Decisions with Multiple Objectives*. Cambridge, UK: Cambridge University Press, 1993.

Kellogg, W. K., *Logic Model Development Guide*, Michigan: WK Kellogg Foundation, 2004.

Kolb, D., and R. Fry, "Towards an Applied Theory of Experiential Learning," in C. L. Copper, ed., *Theories of Group Processes*, London: John Wiley, 1975, pp. 33–58.

Kousky, Carolyn, and Stephen H. Schneider, "Global Climate Policy: Will Cities Lead the Way?" *Climate Policy*, Vol. 3, No. 4, 2003, pp. 359–372.

Krause, R. M., "An Assessment of the Impact that Participation in Local Climate Networks Has on Cities' Implementation of Climate, Energy, and Transportation Policies," *Review of Policy Research*, Vol. 29, No. 5, 2012, pp. 585–604.

Kwakkel, J., W. Walker, and V. Marchau, "Classifying and Communicating Uncertainties in Model-Based Policy Analysis," *International Journal of Technology Management*, Vol. 10, April 2010, pp. 1468–4322.

Lamhauge, Nicolina, Elisa Lanzi, and Shardul Agrawala, "Monitoring and Evaluation for Adaptation: Lessons from Development Co-operation Agencies," Paris: OECD Publishing, OECD Environment Working Papers, No. 38, 2012.

Leagnavar, P., D. Bours, and C. McGinn, *Good Practice Study on Principles for Indicator Development, Selection, and Use in Climate Change Adaptation Monitoring and Evaluation*, Washington, D.C.: Climate-Eval Community of Practice and the Global Environment Facility's Independent Evaluation Office, 2015.

Leclerc, Liza, *Background Document and Workshop Report: Measuring Progress on Adaptation in Canada*, Natural Resources Canada, 2012.

Legey, L. F. L., and H. F. Kazay, "Fuzzy Robustness Analysis," EUSLFAT Conference, 2001, pp. 71–74. As of October 16, 2016:
http://citeseerx.ist.psu.edu/viewdoc/download?doi=10.1.1.145.3567&rep=rep1&type=pdf

Lehmann, Paul, Miriam Brenck, Oliver Gebhardt, Sven Schaller, and Elisabeth Süßbauer, "Understanding Barriers and Opportunities for Adaptation Planning in Cities," UFZ Discussion Papers, No. 19, 2012.

Lempert, R., "Scenarios That Illuminate Vulnerabilities and Robust Responses," *Climatic Change*, Vol. 117, 2013, pp. 627–646.

Lempert, Robert J., "Embedding (Some) Benefit-Cost Concepts into Decision Support Processes with Deep Uncertainty," *Journal of Benefit-Cost Analysis*, Vol. 5, No. 3, December 2014, pp. 487–514.

Lempert, R., and D. G. Groves, "Identifying and Evaluating Robust Adaptive Policy Responses to Climate Change for Water Management Agencies in the American West," *Technological Forecasting and Social Change*, Vol. 77, 2010, pp. 960–974.

Lempert, R. J., D. G. Groves, S. W. Popper, and S. C. Bankes, "A General, Analytic Method for Generating Robust Strategies and Narrative Scenarios," *Management Science*, Vol. 52, No. 4, 2006, pp. 514–528.

Lempert, R., and S. McKay, "Some Thoughts on the Role of Robust Control Theory in Climate-Related Decision Support," *Climatic Change*, Vol. 107, No. 3, 2011, pp. 241–246.

Lempert, R., N. Nakicenovic, D. Sarewitz, and M. Schlesinger, "Characterizing Climate-Change Uncertainties for Decision-Makers—An Editorial Essay," *Climatic Change*, Vol. 65, No. 1–2, 2004, pp. 1–9.

Lempert, Robert J., Steven W. Popper, and Steven C. Bankes, *Shaping the Next One Hundred Years: New Methods for Quantitative, Long-Term Policy Analysis*, Santa Monica, Calif.: RAND Corporation, MR-1626-RPC, 2003. As of October 16, 2016:
http://www.rand.org/pubs/monograph_reports/MR1626.html

Lempert, Robert J., Steven W. Popper, David G. Groves, Nidhi Kalra, Jordan R. Fischbach, Steven C. Bankes, Benjamin P. Bryant, Myles T. Collins, Klaus Keller, Andrew Hackbarth, Lloyd Dixon, Tom LaTourrette, Robert T. Reville, Jim W. Hall, Christophe Mijere, and David J. McInerney, "Making Good Decisions Without Predictions: Robust Decision Making for Planning Under Deep Uncertainty," Santa Monica, Calif.: RAND Corporation, RB-9701, 2013. As of October 16, 2016:
http://www.rand.org/pubs/research_briefs/RB9701.html

Local and Regional Partnership Board, *Adapting to Climate Change: Guidance Notes for National Indicator 188*, Government of the United Kingdom, 2010.

March, J., L. Sproul, and M. Tamuz, "Learning from Samples of One or Fewer," *Organizational Science*, Vol. 2, 1991, pp. 1–13.

Martin-Breen, Patrick, and J. Marty Anderies, *Resilience: A Literature Review*, Rockefeller Foundation, September 18, 2011.

Maxwell, J. A., *Qualitative Research Design: An Interactive Approach*, Thousand Oaks, Calif.: Sage Publications, Inc., 2013.

McCray, L. E., K. A. Oye, and A. C. Petersen, "Planned Adaptation in Risk Regulation: An Initial Survey of U.S. Environmental, Health, and Safety Regulation," *Technological Forecasting and Social Change*, Vol. 77, 2010, pp. 951–959.

Merriam-Webster, "Framework," undated. As of October 16, 2016:
http://www.merriam-webster.com/dictionary/framework

Miles, M. B., and A. M. Huberman, *Qualitative Data Analysis*, 2nd edition, Thousand Oaks, Calif.: Sage Publications, Inc., 1994.

Mimura, N., R. S. Pulwarty, D. M. Duc, I. Elshinnawy, M. H. Redsteer, H. Q. Huang, J. N. Nkem, and R. A. Sanchez Rodriguez, "Adaptation Planning and Implementation," *Climate Change 2014: Impacts, Adaptation, and Vulnerability. Part A: Global and Sectoral Aspects. Contribution of Working Group II to the Fifth Assessment Report of the Intergovernmental Panel on Climate Change*, Cambridge, UK, and New York: Cambridge University Press, 2014, pp. 869–898.

Moench, M., "Experiences Applying the Climate Resilience Framework: Linking Theory with Practice," *Development in Practice*, Vol. 24, No. 4, 2014, pp. 447–464.

Moench, Marcus, Stephen Tyler, and Jessica Lage, *Catalyzing Urban Climate Resilience: Applying Resilience Concepts to Planning Practice in the ACCCRN Program (2009–2011)*, International Institute for Social and Environmental Transition, 2011.

Morgan, M. G., and M. Henrion, *Uncertainty: A Guide to Dealing with Uncertainty in Quantitative Risk and Policy Analysis*, Cambridge, UK: Cambridge University Press, 1990.

Moser, Susanne C., "Good Morning, America! The Explosive US Awakening to the Need for Adaptation," 2009, pp. 1–39.

Moser, S., and M. T. Boykoff, *Toward Successful Adaptation: Linking Science and Policy in a Rapidly Changing World*, London: Routledge, 2013.

Moser, S. C., and J. A. Ekstrom, "A Framework to Diagnose Barriers to Climate Change Adaptation," *Proceedings of the National Academy of Sciences of the United States of America*, Vol. 107, No. 51, 2010, pp. 22026–22031.

Moss, R., P. L. Scarlett, M. A. Kenney, H. Kunreuther, R. Lempert, J. Manning, B. K. Williams, J. W. Boyd, E. T. Cloyd, L. Kaatz, and L. Patton, "Decision Support: Connecting Science, Risk Perception, and Decisions," in J. M. Melilo, T. Richmond, and G. Yohe, *Climate Change Impacts in the United States: The Third National Climate Assessment*, Washington, D.C.: U.S. Global Change Research Program, 2014, pp. 620–647.

National Accounts of Well-Being, "What Is Well-Being?" 2015. As of November 2, 2015:
http://www.nationalaccountsofwellbeing.org/learn/what-is-well-being.html

National Climate Change Adaptation Research Facility, "What Does Climate Change Mean for Australia?" undated. As of October 26, 2015:
https://www.nccarf.edu.au/content/adaptation

National Implementation Research Network, "Implementation Defined," 2016. As of June 26, 2016:
http://nirn.fpg.unc.edu/learn-implementation/implementation-defined

National Research Council, *Informing Decisions in a Changing Climate*, Washington, D.C.: The National Academies Press, Panel on Strategies and Methods for Climate-Related Decision Support, Committee on the Human Dimensions of Climate Change, Division of Behavioral and Social Sciences and Education, 2009.

National Research Council, *Adapting to the Impacts of Climate Change*, Washington, D.C.: The National Academies Press, 2010.

National Research Council, *America's Climate Choices*, Washington, D.C.: The National Academies Press, 2011.

National Research Council, *Disaster Resilience: A National Imperative*, Washington, D.C.: The National Academies Press, 2012.

Nelson, D. R, W. N. Adger, and K. Brown, "Adaptation to Environmental Change: Contributions of a Resilience Framework," *Annual Review of Environment and Resources*, Vol. 32, No. 1, 2007, p. 395.

New York City, Special Initiative for Rebuilding and Resiliency, *PlanNYC: A Stronger, More Resilient New York*, New York: New York City's Office of Long-Term Planning and Sustainability (OLTPS), 2013. As of June 13, 2013:
http://www.nyc.gov/html/sirr/html/report/report.shtml

Non-State Actor Zone for Climate Action, "NAZCA Captures the Commitments to Climate Action by Companies, Cities, Subnational, Regions, Investors, and Civil Society Organizations," 2016. As of May 3, 2016:
http://climateaction.unfccc.int

Oppenheimer, M., M. Campos, R. Warren, J. Birkmann, G. Luber, B. C. O'Neill, and K. Takahashi, "Emergent Risks and Key Vulnerabilities," in C. B. Field, V. R. Barros, D. J. Dokken, K. J. Mach, M. D. Mastrandrea, T. E. Bilir, M. Chatterjee, K. L. Ebi, Y. O. Estrada, R. C. Genova, B. Girma, E. S. Kissel, A. N. Levy, S. MacCracken, P. R. Mastrandrea, and L. L. White, eds., *Climate Change 2014: Impacts, Adaptation, and Vulnerability. Part A: Global and Sectoral Aspects; Contribution of Working Group II to the Fifth Assessment Report of the Intergovernmental Panel of Climate Change*, Cambridge, UK, and New York: Cambridge University Press, 2014, pp. 1039–1099.

Organisation for Economic Co-operation and Development, "Glossary of Key Terms in Evaluation and Results-Based Management," Paris: DAC Network on Development Evaluation, 2002. As of November 3, 2015:
http://www.oecd.org/dac/2754804.pdf

Pahl-Wostl, C., "A Conceptual Framework for Analysing Adaptive Capacity and Multi-Level Learning Processes in Resource Governance Regimes," *Global Environmental Change*, Vol. 19, 2009, pp. 354–365.

Park, S. E., N. A. Marshall, E. Jakku, A. M. Dowd, S. M. Howden, E. Mendham, and A. Fleming, "Informing Adaptation Responses to Climate Change Through Theories of Transformation," *Global Environmental Change*, Vol. 22, 2012, pp. 115–126.

Parker, A. M., S. Srinivasan, R. J. Lempert, and S. Berry, "Evaluating Simulation-Derived Scenarios for Effective Decision Support," *Technological Forecasting and Social Change*, Vol. 91, 2015, pp. 64–77.

Pelling, M., C. High, J. Dearing, and D. Smith, "Shadow Spaces for Social Learning: A Relational Understanding of Adaptive Capacity to Climate Change Within Organisations," *Environment and Planning A.*, Vol. 40, No. 4, 2008, pp. 867–884.

Peschl, M. F., "Triple-Loop Learning as a Foundation for Profound Change, Individual Cultivation, and Radical Innovation: Construction Processes Beyond Scientific and Rational Knowledge," *Constructivist Foundations*, Vol. 2, No. 2–3, 2007, pp. 136–145.

Preston, B., R. Westaway, and E. Yuen, "Climate Adaptation Planning in Practice: An Evaluation of Adaptation Plans for Three Developed Nations," *Mitigation and Adaptation Strategies for Global Change*, Vol. 16, 2011, pp. 407–438.

Preston, B. L., R. M. Westaway, S. Dessai, and T. Smith, "Are We Adapting to Climate Change? Research and Methods for Evaluating Progress," *89th American Meteorological Society Annual Meeting: Fourth Symposium on Policy and Socio-Economic Research*, Phoenix, Ariz., 2009.

Ranger, N., A. Millner, S. Dietz, S. Fankhauser, A. Lopez, and G. Ruta, "Adaptation in the UK: A Decision Making Process," Granthan/CCEP Policy Brief, 2010.

Reams, Margaret A., Kelsey W. Clinton, and Nina S. N. Lam, "Achievement of Climate Planning Objectives Among U.S. Member Cities of the International Council for Local Environmental Initiatives (ICLEI)," *Low Carbon Economy*, Vol. 3, No. 4, 2012.

Rebuild by Design, home page, 2016. As of May 3, 2016:
http://www.rebuildbydesign.org

Renn, O., *Risk Governance: Coping with Uncertainty in a Complex World*, London: Earthscan, 2008.

Resalliance.org, "Resilience," 2015. As of October 26, 2015:
http://www.resalliance.org/resilience

Robson, C., *Real World Research*, 3rd Ed., Hoboken, N.J.: Wiley, 2011.

Rockefeller Foundation, "100 Resilient Cities," undated. As of April 28, 2015:
http://www.100resilientcities.org/#/-_/

Rodin, J., *The Resilience Dividend: Being Strong in a World Where Things Go Wrong*, New York: Perseus Books Group, 2014.

Rosenhead, J., "Robust Analysis: Keeping Your Options Open," in J. Rosenhead and J. Mingers, *Rational Analysis for a Problematic World Revisited: Problem Structuring Methods for Complexity, Uncertainty, and Conflict*, Chichester, UK: John Wiley & Sons, 2001.

Rosenhead, M. J., M. Elton, and S. K. Gupta, "Robustness and Optimality as Criteria for Strategic Decisions," *Operational Research Quarterly*, Vol. 23, No. 4, 1972, pp. 413–430.

Schneider, S. H., S. Semenov, A. Patwardhan, I. Burton, C. H. D. Magadza, M. Oppenheimer, A. B. Pittock, A. Rahman, J. B. Smith, A. Suarez, and F. Yamin, "Assessing Key Vulnerabilities and the Risk from Climate Change," in M. L. Parry, O. F. Canziani, J. P. Palutikof, P. J. van der Linden, and C. E. Hanson, eds., *Climate Change 2007: Impacts, Adaptation and Vulnerability—Contribution of Working Group II to the Fourth Assessment Report of the Intergovernmental Panel on Climate Change*, Cambridge, UK: Cambridge University Press, 2007, pp. 779–810.

Schröter, D., C. Polsky, and A. G. Patt, "Assessing Vulnerabilities to the Effects of Global Change: An Eight Step Approach," *Mitigation and Adaptation Strategies for Global Change*, Vol. 10, No. 4, 2005, pp. 573–595.

Sen, A., *The Idea of Justice*, Cambridge, Mass.: Belknap Press, 2009.

Seto, K. C., S. Dhakal, A. Bigio, H. Blanco, G. C. Delgado, D. Dewar, L. Huang, A. Inaba, A. Kansal, S. Lwasa, J. E. McMahon, D. B. Müller, J. Murakami, H. Nagendra, and A. Ramaswami, "Human Settlements, Infrastructure and Spatial Planning," in O. Edenhofer, R. Pichs-Madruga, Y. Sokona, E. Farahani, S. Kadner, K. Seyboth, A. Adler, I. Baum, S. Brunner, P. Eickemeier, B. Kriemann, J. Savolainen, S. Schlömer, C. von Stechow, T. Zwickel, and J. C. Minx, eds., *Climate Change 2014: Mitigation of Climate Change—Contribution of Working Group III to the Fifth Assessment Report of the Intergovernmental Panel on Climate Change*, Cambridge, UK, and New York: Cambridge University Press, 2014.

Sharp, Elaine B., Dorothy M. Daley, and Michael S. Lynch, "Understanding Local Adoption and Implementation of Climate Change Mitigation Policy," *Urban Affairs Review*, Vol. 47, No. 3, 2011, pp. 433–457.

Singh, Neelam, and Marion Vieweg, "Monitoring Implementation and Effects of GHG Mitigation Policies: Steps to Develop Performance Indicators," Working Paper, World Resources Institute, 2015.

Sippel, Maike, and Till Jenssen, *What About Local Climate Governance? A Review of Promise and Problems*, Social Science Research Network, November 2009.

Southeast Florida Regional Climate Change Compact, "About the Compact," 2016. As of May 3, 2016:
http://www.southeastfloridaclimatecompact.org/who-we-are/

Spearman, Margaret, and Heather McGray, *Making Adaptation Count: Concepts and Options for Monitoring and Evaluation of Climate Change Adaptation*, Washington, D.C.: World Resources Institute, October 2011.

State of California, Global Warming Solutions Act, Assembly Bill (AB) 32, 2006. As of October 17, 2016:
http://www.arb.ca.gov/cc/docs/ab32text.pdf

State of California, Sustainable Communities and Climate Protection Act of 2008, Senate Bill (SB) 375, 2008. As of October 17, 2016:
http://www.leginfo.ca.gov/pub/07-08/bill/sen/
sb_0351-0400/sb_375_bill_20080930_chaptered.pdf

State of California, Executive Order B-30-15, April 29, 2015.

Sterman, J. D., "Communicating Climate Change Risks in a Skeptical World," *Climatic Change*, Vol. 108, 2011, pp. 811–826.

Stern, Nicholas, *The Economics of Climate Change: The Stern Review*, Cambridge, UK, and New York: Cambridge University Press, 2007.

Sussman, Fran, Nisha Krishnan, Kathryn Maher, Rawlings Miller, Charlotte Mack, Paul Stewart, and Bill Perkins, "Climate Change Adaptation Cost in the US: What Do We Know?" *Climate Policy*, Vol. 14, No. 2, 2014, pp. 242–282.

Swart, Rob, Robbert Biesbroek, Svend Binnerup, Timothy R. Carter, Caroline Cowan, Thomas Henrichs, and Daniela Rey, *Europe Adapts to Climate Change: Comparing National Adaptation Strategies*, 2009. As of October 17, 2016:
http://www.peer.eu/fileadmin/user_upload/publications/PEER_Report1.pdf

Swanson, D., S. Barg, S. Tyler, H. Venema, S. Tomar, S. Bhadwal, S. Nair, D. Roy, and J. Drexhage, "Seven Tools for Creating Adaptive Policies," *Technological Forecasting and Social Change*, Vol. 77, 2010, pp. 924–939.

Swanson, D., H. Venema, S. Barg, S. Tyler, J. Drexage, P. Bhandari, and U. Kelkar, *Initial Conceptual Framework and Literature Review for Understanding Adaptive Policies*, 2007.

Swart, R., M. Berk, E. Kreileman, M. Janssen, J. Bollen, R. Leemans, and B. de Vries, *The Safe Landing Approach: Risks and Trade-Offs in Climate Change*, RIVM, 1998.

Tang, Z. H., S. D. Brody, C. Quinn, L. Chang, and T. Wei, "Moving from Agenda to Action: Evaluating Local Climate Change Action Plans," *Journal of Environmental Planning and Management*, Vol. 53, No. 1, 2010, pp. 41–62.

Tingstad, A. H., D. G. Groves, and R. J. Lempert, "Tree-Ring Based Climate Scenarios to Inform Decision Making in Water Resource Management: A Case Study from the Inland Empire, CA," *Journal of Water Resource Planning and Management*, August 2013.

Toth, F. L., "Climate Policy in Light of Climate Science: The ICLIPS Project," *Climatic Change*, Vol. 56, No. 1–2, 2003, pp. 7–36.

Trigeorgis, L., *Real Options: Managerial Flexibility and Strategy in Resource Allocation*, Cambridge, Mass.: MIT Press, 1996.

Tyler, S., and M. Moench, "A Framework for Urban Climate Resilience," *Climate and Development*, Vol. 4, No. 4, 2012, pp. 311–326.

UNFCCC—*see* United Nations Framework Convention on Climate Change.

United Kingdom, "Policy: Adapting to Climate Change," 2014. As of October 17, 2016:
https://www.gov.uk/government/policies/adapting-to-climate-change/supporting-pages/
adaptation-reporting-power

United Nations Environment Programme, "Climate Change Mitigation," 2015. As of November 2, 2015:
http://www.unep.org/climatechange/mitigation/

United Nations Framework Convention on Climate Change, 15th Conference of the Parties, Section 1 of the Copenhagen Accord, 2009. As of October 17, 2016:
http://unfccc.int/resource/docs/2009/cop15/eng/11a01.pdf

United Nations Framework Convention on Climate Change, "FOCUS: Adaptation," 2014. As of October 26, 2015:
http://unfccc.int/focus/adaptation/items/6999.php

United Nations Framework Convention on Climate Change, 20th Conference of the Parties, Paris Agreement, Intended Nationally Determined Contributions (INDCs) as Communicated by the Parties, 2015. As of October 17, 2016:
http://www4.unfccc.int/submissions/indc/Submission%20Pages/submissions.aspx

United Nations, *World Urbanization Prospects, the 2014 revision*, UN Department of Economic and Social Affairs, Population Division, 2014. As of May 3, 2016:
http://esa.un.org/unpd/wup/CD-ROM/Default.aspx

U.S. Agency for International Development, Office of the Director of U.S. Foreign Assistance, "Glossary of Evaluation Terms," 2009. As of November 2, 2015:
http://pdf.usaid.gov/pdf_docs/Pnado820.pdf

U.S. Bureau of the Census, Criteria available from the Chief, Geography Division, U.S. Bureau of the Census, Washington, D.C. 20233, 1995.

U.S. Conference of Mayors, *Climate Mitigation and Adaptation Actions in America's Cities: A 282-City Survey*, Washington, D.C.: Mayors Climate Protection Center, April 2014.

U.S. Conference of Mayors, *MCP Agreement*, 2005. As of October 17, 2016:
http://www.usmayors.org/climateprotection/documents/mcpagreement.pdf

U.S. Conference of Mayors, "List of Participating Mayors," undated. As of January 28, 2014:
http://www.usmayors.org/climateprotection/list.asp

U.S. Department of Housing and Urban Development, "Natural Disaster Resilience Competition—Fact Sheet," September 2014a. As of October 17, 2016:
http://www.hrpdcva.gov/uploads/docs/4A%20-%20NDRC%20Fact%20Sheet.pdf

U.S. Department of Housing and Urban Development, "HUD Launches $1 Billion Natural Disaster Resilience Competition," September 17, 2014b. As of June 17, 2015:
http://portal.hud.gov/hudportal/HUD?src=/press/press_releases_media_advisories/2014/
HUDNo_14-109

U.S. Global Change Research Program, *National Climate Assessment*, Washington, D.C., 2014.

Viggh, A., P. Leagnavar, D. Bours, and C. McGinn, *Good Practice Study on Principles for Indicator Development, Selection, and Use in Climate Change Adaptation Monitoring and Evaluation*, Climate-Eval Community of Practice, 2015. As of October 17, 2016:
https://www.climate-eval.org/sites/default/files/studies/Good-Practice-Study.pdf

Walker, W. E., S. A. Rahman, and J. Cave, "Adaptive Policies, Policy Analysis, and Policy-Making," *European Journal of Operational Research*, Vol. 128, 2001, pp. 282–289.

Walker, W., and V. Marchau, "Dealing with Uncertainty in Policy Analysis and Policy-Making," *Integrated Assessment*, Vol. 4, No. 1, 2003, pp. 1–4.

Walker, B., C. S. Holling, S. R. Carpenter, and A. Kinzig, "Resilience, Adaptability and Transformability in Social–Ecological Systems," *Ecology and Society*, Vol. 9, No. 2, 2004, p. 5.

Wallis, Allan, Mia Colson, and Kristin Heery, "A Survey of Regional Planning for Climate Adaptation," Washington, D.C.: National Association of Regional Councils, 2011.

Weadapt.org, "Iterative Risk Management," 2014. As of October 26, 2015:
https://www.weadapt.org/knowledge-base/climate-adaptation-training/
module-iterative-risk-management

Wheeler, Stephen M., "State and Municipal Climate Change Plans: The First Generation," *Journal of the American Planning Association*, Vol. 74, No. 4, 2008, pp. 481–496.

The White House, "Planning for Federal Sustainability in the Next Decade," Executive Order 13693, November 23, 2015. As of October 17, 2016:
https://www.whitehouse.gov/the-press-office/2015/11/23/
obama-administration-announces-2016-greenhouse-gas-targets-and

Wood, P. J., "Climate Change and Game Theory," in R. Costanza, K. Limburg, and I. Kubiszewski, eds., *Ecological Economics Reviews*, Vol. 1219, Malden: Wiley-Blackwell, 2011, pp. 153–170.

Woodruff, S. C., and M. Stults, "Numerous Strategies but Limited Implementation Guidance in U.S. Local Adaptation Plans," *Nature Climate Change*, May 2, 2016.

World Bank, *World Development Report 2014: Risk and Opportunity—Managing Risk for Development*, Washington, D.C., 2013.

Yohe, G. W., "The Tolerable Windows Approach: Lessons and Limitations," *Climatic Change*, Vol. 41, No. 3, 1999, pp. 283–295.